SpringerBriefs in Computer Science

SpringerBriefs present concise summaries of cutting-edge research and practical applications across a wide spectrum of fields. Featuring compact volumes of 50 to 125 pages, the series covers a range of content from professional to academic.

Typical topics might include:

- A timely report of state-of-the art analytical techniques
- A bridge between new research results, as published in journal articles, and a contextual literature review
- A snapshot of a hot or emerging topic
- An in-depth case study or clinical example
- A presentation of core concepts that students must understand in order to make independent contributions

Briefs allow authors to present their ideas and readers to absorb them with minimal time investment. Briefs will be published as part of Springer's eBook collection, with millions of users worldwide. In addition, Briefs will be available for individual print and electronic purchase. Briefs are characterized by fast, global electronic dissemination, standard publishing contracts, easy-to-use manuscript preparation and formatting guidelines, and expedited production schedules. We aim for publication 8–12 weeks after acceptance. Both solicited and unsolicited manuscripts are considered for publication in this series.

More information about this series at http://www.springer.com/series/10028

B. R. Wienke • T. R. O'Leary

Understanding Modern Dive Computers and Operation

Protocols, Models, Tests, Data, Risk and Applications

B. R. Wienke
Los Alamos National Laboratory
Los Alamos, NM, USA

T. R. O'Leary
NAUI Technical Diving Operations
Riverview, FL, USA

ISSN 2191-5768   ISSN 2191-5776   (electronic)
SpringerBriefs in Computer Science
ISBN 978-3-319-94053-3   ISBN 978-3-319-94054-0   (eBook)
https://doi.org/10.1007/978-3-319-94054-0

Library of Congress Control Number: 2018951058

This Springer imprint is published by the registered company Springer Nature Switzerland AG.
The registered company address is: Gewerbestrasse 11, 6330 Cham, Switzerland

UNDERSTANDING MODERN DIVE COMPUTERS AND OPERATION
Protocols, Models, Tests, Data, Risk and Applications

by B.R. Wienke and T.R. O'Leary

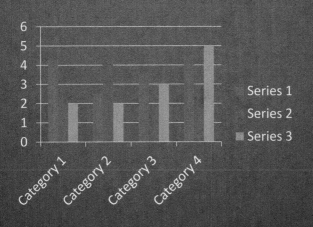

Cover Abstract

This short brief details dive computers, operation, protocols, models, data, tests, risk, and coupled applications. Basic diving principles are detailed with practical computer implementations. Topics are related to diving protocols and operational procedures. Tests and correlations of computer models with data are underscored. The exposition also links phase mechanics to decompression theory with equations used in computer syntheses. Happily today, we are looking at both dissolved gases and bubbles in our staging regimens and not just dissolved gas protocols. Onward through the fog. As research expands, dive computers are quick to incorporate new diving technology and science. References are both extensive and pertinent to topical developments and history. Applications focus upon and mimic dive computer operations within model implementations for added understanding. Intended audience are the computer scientist, doctor, researcher, engineer, physical and life sciences professional, chamber technician, explorer, commercial diver, diving instructor, and technical and recreational divers with a need for a concise yet thorough treatise on dive computers and applications.

Bruce Wienke

- Extreme exposure diver
- C&C Dive Team Leader
- WMD Amelioration
- LANL Program Manager
- Computational Physicist
- Author 12 Books
- USSA Masters Ski Racer
- NAUI Tec/Rec Course Director

Tim O' Leary

- DMT
- CHT
- Sat Diver/Supervisor
- Director NAUI Tec Ops
- CEO American Diving
- Early Deep Stop Testor
- NAUI Training Data Guy
- Worldwide Diver

Preface

This brief focuses on dive computers, operation, protocols, models, data, tests, risk, and associated applications in working detail. Basic diving principles are framed as incorporated and implemented in dive computers. Topics are keyed to diving protocols and operational procedures. Tests and correlations of models with data are underscored. The exposition also links phase mechanics to decompression theory with equations used in computer synthesis. References are both extensive and pertinent to topical development and history. Applications focus upon and mimic dive computer operations and model implementations for added understanding. A recap with questions and answers posed to the Authors completes the brief. The intended audience are the computer scientist, doctor, researcher, engineer, physical and life sciences professional, chamber technician, explorer, commercial diver, diving instructor, and technical and recreational divers with a need for a concise yet thorough treatise on dive computers and applications.

Theory and dive computer application are, at times, more an art form than exact science. Some believe deterministic modeling is only fortuitous. Technological advance, elucidation of competing mechanisms, and resolution of model issues over the past 100 years have not been rapid. Model implementations tend to be ad hoc, tied to data fits and difficult to quantify on just first principles. Almost any description of decompression processes in tissue and blood can be disputed and possibly turned around on itself. The fact that decompression sickness occurs in metabolic and perfused matter makes it difficult to design and analyze experiments outside living matter. Yet, for application to safe diving, we need models, tests, data, and correlations to build tables and dive computers. And, regardless of biological complexity, certain coarse grain biophysics principles, some neglected in the past, are making a substantial change in diver-staging regimens, decompression theory, and coupled data analysis. Happily today, we are looking at both dissolved gases and bubbles in staging regimens and not just dissolved gas approaches of Haldane, As research expands, dive computers will be quick to reflect new diving technology and science. And that means enhanced safety.

Happy and safe diving always.

Acknowledgments

Thanks to our friends and colleagues at LANL, NAUI, C&C Dive Team Operations, DAN, Commercial Diving Industry, recreational and technical training agencies, computer manufacturers worldwide, and many collaborators at universities, national laboratories, DOE, DOD, USAF, USCG, and USN. Special thanks to my beautiful and talented wife, Luzanne Coburn, for artwork, photography, and finishing.

Author Sketches

Bruce Wienke

Bruce Wienke is a program manager in the Weapons Technology/Simulation Office at LANL. He received a BS in physics and mathematics (Northern Michigan), MS in nuclear physics (Marquette) and a PhD in particle physics (Northwestern). He has authored 250+ articles in peer-reviewed journals, media outlets, trade magazines, and workshop proceedings and has published 12 books on diving science, biophysics, and decompression theory. He heads up the C& C Dive Team vested with worldwide underwater search, assessment, and disablement of nuclear, chemical, and biological WMDs. He is a fellow of the APS, technical committee member of the ANS, and member of the UHMS and serves as a consultant to the EPA, DHS, ADA, US military, and dive industry. Bruce is an editor/reviewer for CBM, PR, TTSP, NSE, and JQSRT, and CEO of Southwest Enterprises Consulting. He is the developer of the Reduced Gradient Bubble Model (RGBM) implemented in decompression meters, tables, and dive software worldwide. Bruce has dived all over the world on OC and RB systems in military, scientific, exploration, testing, and training activities. He is a NAUI Tec/Rec instructor trainer and course director. Interests include USSA masters ski racing, USTA seniors tennis, golf, and windsurfing. Bruce is a certified ski instructor (PSIA) and racing coach (USSCA). He has won masters national titles in SL, GS, SG, and DH and quarterbacked the Northern Michigan Wildcats to a NCAA II Title in the Hickory Bowl.

Tim O'Leary

Tim O'Leary heads up NAUI Technical Diving Operations having developed and co-authored training manuals, support material, tech dive tables, monographs and related media along with tech course standards. He is a practicing commercial diver and CEO of American Diving & Marine Salvage on the Texas Gulf Coast. Tim received a BS in zoology (Texas A&M) and a DMT and CHT from Jo Ellen Smith Medical Center at the Baromedical Research Institute. He was a commercial diving and hyperbaric chamber instructor at the Ocean Corporation. Tim is a member of the UHMS, SNAME, and NADMT. He is an admiral in the Texas Navy, a USCG 100 Ton Vessel Master, and a consultant to Texas Parks & Wildlife, Canadian

Corporation, Rimkus Group, and Offshore Oil Industry. His diving experience is global on OC and RB systems in commercial, exploration, training, and testing activities. He is a NAUI Tec/Rec instructor trainer, course director, and workshop director. Other interests include skiing, deep wreck diving, and dive travel. Tim and NAUI Dive Team are credited with the discovery and exploration of the USS Perry in approximately 250 fsw off Anguar and diving it for a week on RBs.

Units and Fundamental Constants

Note so-called diving units are employed herein, that is, standard SI units for depth and pressure are not used. Instead, pressures and depths are both measured in feet-of-seawater (fsw) or meters-of-seawater (msw). The conversion is standard,

$$10\,msw = 33.28\,fsw = 1\,atm$$

Specific densities, η (dimensionless) in pressure relationships, are normalized to sea water density with freshwater-specific density equal to 0.975.

Breathing mixtures, such as nitrox (nitrogen and oxygen), heliox (helium and oxygen), and trimix (helium, nitrogen and oxygen), carry standardized notation. If the fraction of oxygen is greater than 21%, the mixture is termed enriched. Enriched nitrox mixtures are denoted EANx; enriched heliox mixtures are denoted EAHx. For other mixtures of nitrox and heliox, the convention is to name them with inert gas percentage first and then oxygen percentage, such as 85/15 nitrox or 85/15 heliox. For trimix, notation is shortened to list the oxygen percentage first and then only the helium percentage, such as 15/45 trimix, meaning 15% oxygen, 45% helium, and 40% nitrogen. Or TMX 15/45 is used. Air is sometimes noted EAN21 or 79/21 nitrox even though not enriched.

Unit Conversion Table

Time

$$1\,sec = 10^3\,msec = 10^6\,\mu sec = 10^9\,nsec$$
$$1\,megahertz = 10^6\,hertz = 10^6\,sec^{-1}$$

Length

$$1\,m = 3.28\,ft = 1.09\,yd = 39.37\,in$$
$$1\,\mu m = 10^4\,angstrom = 10^3\,nm = 10^{-6}\,m$$
$$1\,km = 0.62\,mile$$
$$1\,fathom = 6\,ft$$
$$1\,nautical\,mile = 6,080\,ft = 1.15\,mile = 1.85\,km$$
$$1\,light\,year = 9.46 \times 10^{12}\,km = 5.88 \times 10^{12}\,mile$$

$$1 \ parsec = 3.086 \times 10^{16} \ m = 3.262 \ light \ year$$

Speed

$$1 \ km/hr = 27.77 \ cm/sec$$
$$1 \ mile/hr = 5280 \ ft/sec$$
$$1 \ knot = 1.15 \ mi/hr = 51.48 \ cm/sec$$

Volume

$$1 \ cm^3 = 0.06 \ in^3$$
$$1 \ m^3 = 35.32 \ ft^3 = 1.31 \ yd^3$$
$$1 \ l = 10^3 \ cm^3 = .04 \ ft^3 = 1.05 \ qt$$

Mass, Density, and Viscosity

$$1 \ kg = 32.27 \ oz = 2.20 \ lb$$
$$1 \ g/cm^3 = 0.57 \ oz/in^3$$
$$1 \ kg/m^3 = 0.06 \ lb/ft^3$$
$$1 \ dyne \ sec/cm^2 = 1 \ poise = 0.10 \ pascal \ sec = 0.01 \ poiseuille$$

Force and Pressure

$$1 \ newton = 10^5 \ dyne = 0.22 \ lb$$
$$1 \ g/cm^2 = 0.23 \ oz/in^2$$
$$1 \ kg/m^2 = 0.20 \ lb/ft^2$$
$$1 \ atm = 32.56 \ fsw = 10 \ msw = 1.03 \ kg/cm^2 = 14.69 \ lbs/in^2$$

Energy and Power

$$1 \ cal = 4.19 \ joule = 3.96 \times 10^{-3} \ btu = 3.09 \ ft \ lb$$
$$1 \ joule = 10^7 \ ergs = 0.74 \ ft \ lb$$
$$1 \ keV = 10^3 \ eV = 1.60 \times 10^{-16} \ joule$$
$$1 \ amu = 931.1 \ MeV$$
$$1 \ watt = 3.41 \ btu/hr = 1.34 \times 10^{-3} \ hp$$

Electricity and Magnetism

$$1 \ coul = 2.99 \times 10^9 \ esu$$
$$1 \ amp = 1 \ coul/sec = 1 \ volt/ohm$$
$$1 \ volt = 1 \ newton \ coul \ m = 1 \ joule/coul$$
$$1 \ gauss = 10^{-4} \ weber/m^2 = 10^{-4} \ newton/amp \ m$$
$$1 \ f = 1 \ coul/volt$$

Fundamental constants are listed next.

Fundamental Constants

$$g_0 = 9.80 \ m/sec^2 \quad (Sea \ Level \ Acceleration \ Of \ Gravity)$$
$$G_0 = 6.67 \times 10^{-11} \ newton \ m^2/kg^2 \quad (Gravitational \ Constant)$$
$$M_0 = 5.98 \times 10^{24} \ kg \quad (Mass \ of \ the \ Earth)$$
$$\Gamma_0 = 1.98 \ cal/min \ cm^2 \quad (Solar \ Constant)$$
$$c = 2.998 \times 10^8 \ m/sec \quad (Speed \ of \ Light)$$
$$h = 6.625 \times 10^{-34} \ joule \ sec \quad (Planck \ Constant)$$
$$R = 8.317 \ joule/gmole \ ^\circ K \quad (Universal \ Gas \ Constant)$$

$$k = 1.38 \times 10^{-23} \, joule/gmole \, °K \quad (Boltzmann \, Constant)$$
$$N_0 = 6.025 \times 10^{23} \, atoms/gmole \quad (Avogadro \, Number)$$
$$m_0 = 9.108 \times 10^{-31} \, kg \quad (Electron \, Mass)$$
$$e_0 = 1.609 \times 10^{-19} \, coulomb \quad (Electron \, Charge)$$
$$r_0 = 0.528 \, angstrom \quad (First \, Bohr \, Orbit)$$
$$\epsilon_0 = (4\pi)^{-1} \times 1.11 \times 10^{-10} \, f/m \quad (Vacuum \, Permittivity)$$
$$\mu_0 = 4\pi \times 10^{-7} \, h/m \quad (Vacuum \, Permeability)$$
$$\kappa_0 = (4\pi\epsilon_0)^{-1} = 8.91 \times 10^9 \, m/f \quad (Coulomb \, Constant)$$
$$\alpha_0 = \mu_0/4\pi = 1 \times 10^{-7} \, h/m \quad (Ampere \, Constant)$$
$$\sigma_0 = 5.67 \times 10^{-8} \, watt/m^2 \, K°4 \quad (Stefan - Boltzmann \, Constant)$$

Metrology is the science of measurement and broadly construed encompasses the bulk of experimental science. In the more restricted sense, metrology refers to maintenance and dissemination of a consistent set of units, support for enforcement of equity in trade by weights, and measure laws and process control for manufacturing.

A measurement is a series of manipulations of physical objects or systems according to experimental protocols producing a number. The objects or systems involved are test objects, measuring devices, or computational operations. The objects and devices exist in and are influenced by some environment. The number relates to the some unique feature of the object, such as the magnitude, or the intensity, or the weight or time duration. The number is acquired to form the basis of decisions effecting some human feature or goal depending on the test object.

In order to solidify metrics for useful decision, metrology requires that any number obtained is functionally identical whenever and wherever the measurement process is performed. Such a universally reproducible measurement is called a *proper measurement* and leads to describing *proper quantities*. The equivalences above relate *proper quantities* to the fundamental constants following and permit closure of physical laws. Unit conversion follows via the chain rule, where the unit identities in the table define equivalence ratios that work like simple arithmetic fractions. Units cancel just like numbers. For instance, from the first table,

$$10 \, l = \frac{10^3 \, cm^3}{1 \, l} \times 10 \, l = 10^4 \, cm^3$$

Acronyms and Definitions

Acronyms are useful and standard throughout the dive community and are employed herein:

ANDI: Association of Nitrox Diving Instructors

BM: Bubble phase model dividing the body into tissue compartments with halftimes that are coupled to inert gas diffusion across bubble film surfaces of exponential size distribution constrained in cumulative growth by a volume limit point

Bubble broadening: Noted laboratory effect that small bubbles increase and large bubbles decrease in number in liquid and solid systems due to concentration gradients that drive material from smaller bubbles to larger bubbles over time spans of hours to days

Bubble regeneration: Noted laboratory effect that pressurized distributions of bubbles in aqueous systems return to their original non-pressurized distributions in time spans of hours to days

CCR: Closed-circuit rebreather, a special RB system that allows the diver to fix the oxygen partial pressure in the breathing loop (setpoint)

CMAS: Confederation Mondiale des Activites Subaquatiques

Critical radius: Temporary bubble radius at equilibrium, that is, pressure inside the bubble just equals the sum of external ambient pressure and film surface tension

DB: Data bank storing downloaded computer profiles in 5–10 sec time-depth intervals

DCS: Crippling malady resulting from bubble formation and tissue damage in divers breathing compressed gases at depth and ascending too rapidly

Decompression stop: Necessary pause in a diver ascent strategy to eliminate dissolved gas and/or bubbles safely and is model based with stops usually made in 10 fsw increments

Deep stop: Decompression stop made in the deep zone to control bubble growth

DAN: Divers Alert Network

Diveware: Diver staging software package usually based on USN, ZHL, VPM, and RGBM algorithms mainly

Diluent: Any mixed gas combination used with pure oxygen in the breathing loop of RBs

Diving algorithm: Combination of a gas transport and/or bubble model with coupled diver ascent strategy

DOD: Department of Defense

DOE: Department of Energy

Doppler: A device for counting bubbles in flowing blood that bounces acoustical signals off bubbles and measures change in frequency

DSAT: Diving Science and Technology, a research arm of PADI

DSL: Diving Safety Laboratory, the European arm of DAN

EAHx: Enriched air helium breathing mixture with oxygen fraction, x, above 21% often called helitrox

EANx: Enriched air nitrox breathing mixture with oxygen fraction, x, above 21%

EOD: End of dive risk estimator computed after finishing dive and surfacing

ERDI: Emergency Response Diving International

FDF: Finnish Diving Federation

GF: Gradient factor, multiplier of USN and ZHL critical gradients, G and H, that try to mimic BMs

GM: Dissolved gas model dividing the body into tissue compartments with arbitrary half times for uptake and elimination of inert gases with tissue tensions constrained by limit points

GUE: Global Underwater Explorers

Heliox: Breathing gas mixture of helium and oxygen used in deep and decompression diving

IANTD: International Association of Nitrox and Technical Divers

ICD: Isobaric counter diffusion, inert dissolved gases (helium, nitrogen) moving in opposite directions in tissue and blood

IDF: Irish Diving Federation

LSW theory: Lifschitz-Slyasov-Wagner Ostwald bubble ripening theory and model

M-values: Set of limiting tensions for dissolved gas buildup in tissue compartments at depth

Mixed gases: Combination of oxygen, nitrogen, and helium gas mixtures breathed underwater

NAUI: National Association of Underwater Instructors

NDL: No decompression limit, maximum allowable time at given depth permitting direct ascent to the surface

NEDU: Naval Experimental Diving Unit, diver testing arm of the USN in Panama City

Nitrox: Breathing gas mixture of nitrogen and oxygen used in recreational diving

OC: Open circuit, underwater breathing system using mixed gases from a tank exhausted upon exhalation

Ostwald ripening: Large bubble growth at the expense of small bubbles in liquid and solid systems

OT: Oxtox, pulmonary and/or central nervous system oxygen toxicity resulting from over exposure to oxygen at depth or high pressure

PADI: Professional Association of Diving Instructors

PDE: Project Dive Exploration, a computer dive profile collection project at DAN

Phase volume: Surfacing limit point for bubble growth under decompression

RB: Rebreather, underwater breathing system using mixed gases from a cannister that are recirculated after carbon dioxide is scrubbed with oxygen from another cannister injected into the breathing loop

Recreational diving: Air and nitrox nonstop diving

RGBM algorithm: An American bubble staging model correlated with DCS computer outcomes by Wienke

RN: Royal Navy

SDI: Scuba Diving International

Shallow stop: Decompression stop made in the shallow zone to eliminate dissolved gas

SI: Surface interval, time between dives

SSI: Scuba Schools International

TDI: Technical Diving International

Technical diving: Mixed gas (nitrogen, helium, oxygen), OC and RB, deep and decompression diving

TMX x/y: Trimix with oxygen fraction, x, helium fraction, y, and the rest nitrogen

Trimix: Breathing gas mixture of helium, nitrogen, and oxygen used in deep and decompression diving

USAF: United States Air Force

USCG: United States Coast Guard

USN: United States Navy

USN algorithm: An American dissolved gas staging model developed by Workman of the US Navy

UTC: United Technologies Center, an Israeli company marketing a message sending-receiving underwater computer system (UDI) using sonar, GPS, and underwater communications with range about 2 *miles*

VPM algorithm: An American bubble staging model based on gels by Yount

Z-values: Another set of limiting tensions extended to altitude and similar to M-values

ZHL algorithm: A Swiss dissolved gas staging model developed and tested at altitude by Buhlmann

Contents

List of Tables

List of Figures

Diving and Dive Computer History

Dive computers are useful tools across recreational and technical diving [1–6]. They are supplanting traditional dive tables and their use is growing as diving research advances [7–13]. Able to process depth-time readings in fractions of a second, modern dive computers routinely estimate hypothetical dissolved gas loadings, bubble buildup, ascent and descent rates, diver ceilings, time remaining, decompression staging, oxygen toxicity and many related variables. Estimates of these parameters made during or after the dive rely on two basic approaches [4, 6, 8], namely, the classical dissolved gas model (GM) and the modern bubble phase model (BM). Both have seen meaningful correlations with real diving data over limited ranges but differ in staging regimens. Dissolved gas models (GM) focus on controlling and eliminating hypothetical dissolved gas by bringing the diver as close to the surface as possible. Bubble phase models (BM) focus on controlling hypothetical bubble growth and coupled dissolved gas by staging the diver deeper before surfacing. The former gives rise to *shallow* decompression stops while the latter requires *deep* decompression stops in the popular lingo these days. As models go both are fairly primitive only addressing the coarser dynamics of dissolved gas buildup and bubble growth in tissues and blood. Obviously, their use and implementation is limited, but purposeful and useful when correlated with available data and implemented in dive computers. To coin a phrase from a community at large, *all models are wrong but some are useful*. Useful ones are models correlated with diving data, that is, test, databank or safe operational record. As research plods forward, computer Manufacturers are usually quick and flexible in responding to technology updates thereby adding to computer viability as a diving tool. It's reasonable to expect computer usage in diving to grow with commensurate sophistication.

Today, some 15–25 companies manufacture dive computers employing both the GM and BM in another 50–70 models by last count. Recreational dive computers mainly rely on the GM while technical dive computers use the BM. In the limit of nominal exposures and short time (nonstop diving), the GM and BM converge

B. R. Wienke, T. R. O'Leary, *Understanding Modern Dive Computers and Operation*, SpringerBriefs in Computer Science, https://doi.org/10.1007/978-3-319-94054-0_1

in diver staging. Dive planning and decompression software are usually available from computer Vendors. Presently, non offer risk estimations, on-the-fly (OTF) or end-of-dive (EOD). They are needed. We include a general method across OC and RB diving to estimate risk at any point on a dive using exponential supersaturation risk functions correlated with diver profile and DCS outcomes from computer downloaded records in the LANL DB. We link dive computers, computational models, statistical techniques, data and related protocols with risk estimators and applications.

Decompression theory and application are not first principle science. Some believe deterministic modeling is a waste of time. Technological advance, elucidation of competing mechanisms and resolution of model issues over the past 90 years has been slow. Model applications tend to be *ad hoc*, loosely tied to data fits and difficult to quantify on just first principles. Decompression sickness occurs in metabolic and perfused matter making it almost impossible to design and analyze experiments outside living matter. Yet, for application to safe diving, we need models to build dive computers, software and tables. And regardless of biological complexity certain coarse grain physics principles, often neglected in the past, are making a substantial change in diver staging regimens, decompression theory and coupled data analyses. Happily today, we are looking at both dissolved gases and bubbles in our staging regimens. And that is a good thing. Both work for different reasons and safely judging from the collective record of both dissolved gas and bubble based computers.

Applications focus upon and mimic dive computer operations and model implementations for added understanding. Intended audience is the computer scientist, doctor, researcher, engineer, physical and life sciences professional, chamber technician, explorer, commercial diver, diving instructor, technical and recreational diver with a need for a concise yet thorough treatise on dive computers and applications. We hope it meets expectations.

A little diving and computer history is interesting before jumping into modern dive computers, protocols, models, tests, data, risk, applications and interplay. Diving has come a long way over centuries.

Early Diving

Breathold diving dates as far back as Xerxes (519 BC) and Alexander the Great (356 BC) as chronicled earliest by the Greek historian, Herodotus, some 500 years before Christ. Reed breathing tubes were employed by ambushing Roman Legions, and primitive diving bells also date BC. The inverted receptacles utilizing Boyle's law compressed air as breathing mixture gained renown in the 1600s. Persians also carried air in goat skins diving underwater. Today, some Korean and Japanese breathold divers still gather pearls and sponges in the same old way. Combatants in the Mekong Delta of Vietnam used reed breathing tubes in operations in the 1960s.

Surface supplied air and demand regulators were employed in hard hat diving by the 1800s with the first demand regulator, patented by Rouquayrol in 1866, supplied by hand bellows. The first case of nitrogen narcosis was reported by Junod in 1835. Full diving suits in which air escapes through a one way exhaust valve were invented in 1840 and a few are still around. In 1878, Fleuss built the first RB system connecting a face mask and breathing bag to a copper tank that held oxygen and cannister filled with potash to absorb carbon dioxide. Quietly, the revolutionary *aqua lung* of Cousteau, a refinement of the Rouquayrol surface supplied demand regulator, ushered the modern era of SCUBA in wartime Europe in 1943. Diving would never be the same afterward. Freed from surface umbilical, the Cousteau unit provided an element of stealth to tactical underwater operations to be sure but the impact on non military diving was orders of magnitude greater. Coupled to high pressure compressed air in tanks, SCUBA offered the means to explore the underwater world for fun and profit.

Commercial availability of the demand regulator in 1947 initiated sport diving and a fledgling equipment industry. Serious diver training and certifying Agencies, such as the YMCA, National Association of Underwater Instructors (NAUI) and Professional Association of Diving Instructors (PADI), strong and still vital today, organized in the late 1950s and 1960s. In the mid 1950s, the Royal Navy (RN) released their bulk diffusion decompression tables while a little later, in 1958, the US Navy (US) compiled their modified Haldane tables with six perfusion limited compartments. Both would acquire biblical status over the next 25 years, or so. In the mid to late 1950s, Fredrickson in the USA and Alinari in Italy designed and released the first analog decompression meters or computers emulating tissue gas uptake and elimination with pressure gauges, porous plugs and distensible gas bags. The first digital computers, designed by DCIEM in Canada, appeared in the mid 1950s. Employed by the Canadian Navy (CN), they were based on a four compartment (pneumatic) analog model of Kidd and Stubbs. Following introduction of a twelve compartment Haldanian device by Barshinger and Huggins in 1983 decompression computers reached a point of utility and acceptance. Flexible, more reliable to use and able to emulate pertinent mathematical models, digital computers rapidly replaced pneumatic devices in the 1980s. Their timely functionality and widespread use heralded the present era of hi-tech diving, with requirements for comprehensive decompression models across a full spectrum of activity. Computer usage statistics, gathered in the 1990s, point to an enviable track record of diver safety with an underlying decompression sickness (DCS) incidence below 0.10% roughly.

Diver mobility concerns ultimately fostered development of the modern SCUBA unit and the quest to go deeper led to exotic gas breathing mixtures. High pressure cylinders and compressors similarly expedited deeper diving and prolonged exposure time. The world record dives of Keller to 1,000 feet of sea water (fsw) in 1960 not only popularized multiple gas mixtures but also witnessed the first use of computers to generate decompression schedules. Saturation diving and underwater habitats followed soon after spurred by a world thirst for oil. Both multiple gas mixtures and saturation diving became a way of life for some commercial divers by the 1970s, particularly after the oil embargo. Oil concerns still drive the commercial diving industry today.

Cochrane in England invented the high pressure caisson in 1830. Shortly afterward, the first use of a caisson in 1841 in France by Triger also witnessed the first case of decompression sickness, aptly termed the bends because of the position assumed by victims to alleviate the pain. Some fifty years later, in 1889, the first medical lock was employed by Moir to treat bends during construction of the Hudson River Tunnel. Since that time many divers and caisson workers have been treated in hyperbaric chambers. Indeed, the operational requirements of diving over the years have provided the incentives to study hyperbaric physiology and its relationship to decompression sickness and impetus for describing fundamental biophysics. Similarly, limitations of nitrogen mixtures at depth, because of narcotic reactivity, prompted recent study and application of helium, nitrogen, hydrogen and oxygen breathing mixtures at depth, especially in the commercial sector. Today, technical divers on mixed gases ply optimal and safe ascent procedures on OC and RB systems.

Increases in pressure with increasing depth underwater impose many of the limitations in diving, applying equally well to the design of equipment used in this environment. Early divers relied on their breathholding ability while later divers used diving bells. Surface supplied air and SCUBA are rather recent innovations. With increasing depth and exposure time, divers encountered a number of physiological and medical problems constraining activity with decompression sickness perhaps the most restrictive. By the 1800s, bubbles were noted in animals subject to pressure reduction. In the 1900s, they were postulated as the cause of DCS in caisson workers and divers. Within that postulate and driven by a need to both optimize diver safety and time underwater, decompression modeling consolidated early rudimentary schedules into present more sophisticated tables and meters. As knowledge and understanding of DCS increase, so should the validity, reliability and range of applicability of models.

Modern Diving

A consensus of opinions and for a variety of reasons suggests that modern diving began in the early 1960s. Technological achievements, laboratory programs, military priorities, safety concerns, commercial diving requirements and international business spurred diving activity and scope of operation. Diving bells, hot water heating, mixed gases, saturation, deep diving, expanded wet testing, computers and need for efficient decompression algorithms signaled the modern diving era. Equipment advances in open (OC) and closed circuit (RB) breathing devices, wet and dry suits, gear weight, mask and fin design, high pressure compressors, flotation and buoyancy control vests, communications links, gauges and meters, lights, underwater tools (cutting, welding, drilling, explosives), surface supplied air and photographic systems paced technological advances. Training and certification requirements for divers in military, commercial, sport and scientific sectors took definition with growing concern for underwater safety and well being.

In the conquest and exploration of the oceans, saturation diving gained prominence in the 1960s, permitting exploitation of the continental shelf impossible with the short exposure times allowed by conventional diving. Spurred by both industrial and military interests in the ability of men to work underwater for long periods of time notable *habitat* experiments, such as Sealab, Conshelf, Man In Sea, Gulf Task, Tektite and Diogene, demonstrated the feasibility of living and working underwater for long periods of time. These efforts followed proof of principle validation by Bond and coworkers (USN) in 1958 of saturation diving. Saturation exposure programs and tests have been conducted from 35 to 2,000 fsw.

The development and use of underwater support platforms such as habitats, bell diving systems, lockout, free flooded submersibles and diver propulsion units also accelerated in the 1960s and 1970s for reasons of science and economics. Support platforms extended both diver usefulness and bottom time by permitting him to live underwater, reducing descent and ascent time, expanding mobility and lessening physical activity. Today, themselves operating from underwater platforms, remotely operated vehicles (ROV) scan the ocean depths at 6,000 fsw for minerals and oil.

Around 1972, strategies for diving in excess of 1,000 fsw received serious scrutiny, driven by a commercial quest for oil and petroleum products and the needs of the commercial diving industry to service that quest. Questions concerning pharmacological additives, absolute pressure limits, thermal exchange, therapy, compression-decompression procedures, effective combinations of mixed breathing gases and equipment pressure functionality addressed many fundamental issues, unknown or only partially understood. By the early 1980s, it became clear that open sea water work in the 1,000 to 2,000 fsw range was entirely practical and many of the problems, at least from an operational point of view, could be solved. Today, the need for continued deep diving remains with demands that cannot be answered with remote or 1 atm diver systems. Heliox and trimix have become standards for deep excursion breathing gases with trimix the choice for shallower exposures and trimix and heliox the choices for deeper exposures in the field. The original Haldane dissolved gas approach to staging (bringing the diver close to the surface) has given room to deep stops in technical and scientific diving communities. Bubble models are supplanting dissolved gas models for staging and decompression meters and deep stop tables have been field tested and used since the late 1990s with notable success and safety records. Data collection efforts were initiated on recreational and technical fronts in the mid 1990s, notably Project Dive Exploration (PDE) at DAN and LANL Data Bank at LANL. Rebreathers, particularly closed circuit rebreathers (CCR) with fixed oxygen partial pressure, (pp_{O_2}), setpoints, are permitting divers from reports to operate in the 400 fsw range comfortably with both shallow and deep stop technology. Coupled to rapid growth in technical diving activities, computer software and special tables are driving an evolution to revolution in modern diving, particularly since the early 1990s. Software incorporating both dissolved gas (USN Workman, Swiss Buhlmann, US Spencer and Canadian DCIEM to name a few) and bubble (Wienke RGBM and Yount VPM) models became accessible starting in the mid 1980s. Decompression meters using the Haldane approach to staging also came online in the same time frame. Full bubble model implementations in decompression meters were released by Suunto, Mares, Dacor, Hydrospace, Atomic Aquatics, Liquivision, VR Technology, UTC and others by the late 1990s. Many more are on the market today.

Yet, despite tremendous advances in deep diving technology most of the ocean floor is outside human reach. Breathing mixtures that are compressible are limiting. Breathing mixtures that are not compressible offer interesting alternatives. In the 1960s, serious attention was given to liquid breathing mixtures, physiological saline solutions. Acting as inert respiratory gas diluents, oxygenated fluids have been used as breathing mixtures thereby eliminating decompression requirements. Some synthetic fluids, such as fluorocarbon, (FX_{80}), exhibit enormous oxygen dissolution properties. Work continues on these and other pharmacological agents.

On the dive computer side, technological advances on processor raw speed, expanded storage capabilities, oxygen and pressure sensors, wi-fi communications, color displays, user interfaces, profile and information PC interfaces, coupled dive planning software and safety protocols have elevated diving and operational underwater activities to new heights. Details follow.

References

1. Brubakk AO and Neuman TS, *Physiology and Medicine of Diving*, Saunders Publishing, 2003, London.
2. Hills BA, *Decompression Sickness: The Biophysical Basis of Prevention and Treatment*, Wiley and Sons Publishing, 1977, Bath.
3. Bove AA and Davis JC, *Diving Medicine*, Saunders Publishing, 2004, Philadelphia.
4. Wienke BR, *Science of Diving*, CRC Press, 2015, Boca Raton.
5. Joiner JJ, *NOAA Diving Manual: Diving for Science and Technology*, Best Publishing, 2001, Flagstaff.
6. Wienke BR, *Biophysics and Diving Decompression Phenomenology*, Bentham Science Publishers, 2016, Sharjah.
7. Schreiner HR and Hamilton RW, *Validation of Decompression Tables*, Undersea and Hyperbaric Medical Society Workshop, 1989, UHMS Publication 74(VAL) 1–1-88, Washington DC.
8. Wienke BR, *Basic Decompression Theory and Application*, Best Publishing, 2003, San Pedro.
9. Wienke BR and Graver DL, *High Altitude Diving*, NAUI Technical Publication, 1991, Montclair.
10. Blogg SL, Lang MA and Mollerlokken A, *Validation of Dive Computers Workshop*, 2011, EUBS/NTNU Proceedings, Gdansk.
11. Lang MA and Hamilton RW, *AAUS Dive Computer Workshop*, University of Southern California Sea Grant Publication, 1989, USCSG-TR-01–89; Los Angeles.
12. Vann RD, Dovenbarger J and Wachholz, *Decompression Sickness in Dive Computer and Table Use*, DAN Newsletter 1989; 3–6.
13. Westerfield RD, Bennett PB, Mebane Y and Orr D, *Dive Computer Safety*. Alert Diver 1994; Mar-Apr: 1–47.

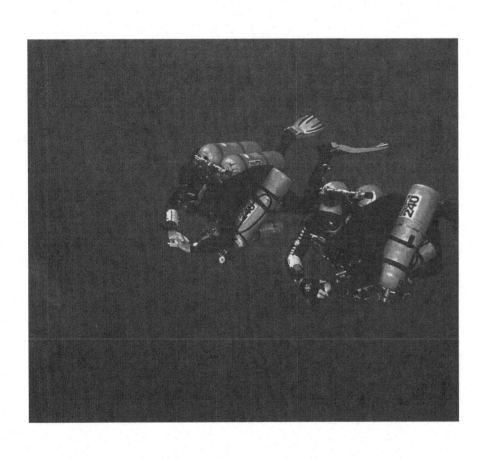

Modern Dive Computers

On the heels of growing interest in underwater science and exploration following World War II, monitoring devices have been constructed to control diver exposure and decompression procedures [1–14]. Devices with records of varying success include mechanical and electrical analogs and within the past 15 years microprocessor based digital computers. With inexpensive microprocessor technology, recent years have witnessed explosive growth in compact digital meter usage. Many use the simple dissolved tissue gas model proposed by Haldane some 100 years ago, but given the sophistication of these devices, most feel that broader models can be incorporated into meter function today increasing range and flexibility. Although the biophysics of bubble formation, free and dissolved phase buildup and elimination is formidable and not fully understood, contemporary models treating both dissolved and free phases, correlated with existing data and consistent with diving protocols extend the utility of diving computers. In the industry, such new models are also termed phase models because they focus on both dissolved gas and bubbles (dual phase).

Statistics point to an enviable track record of decompression meter usage in nominal diving activities as well as an expanding user community. When coupled to slow ascent rates and safety stops, computer usage has witnessed a very low incidence rate of decompression sickness, below 0.01% according to some reports. Computers for nitrox are presently online today along with heliox and trimix units a rather simple code extension of software on any nitrox unit using existing decompression models. Technical divers on mixed gases and making deep decompression dives on OC and RB systems use modern dive computers based on dissolved gas and bubble models for their activities. The modern technical diver also carries wrist slates for decompression schedules extracted from tables and software and blended with a particular brand of personal safety gained from knowledge and experience. Deep stops are an integral part of their diving activities, whether bubble models propose them exactly or are juxta positioned on their ascent profiles by diver choices. These computer units are not inexpensive but their use is expanding

B. R. Wienke, T. R. O'Leary, *Understanding Modern Dive Computers and Operation*, SpringerBriefs in Computer Science, https://doi.org/10.1007/978-3-319-94054-0_2

across both technical and recreational diving. So is decompression diving across all sectors of exploration, scientific, military and related activities. Old recreational prohibitions against decompression diving by Training Agencies has given way to technical programs for mixed gas, OC and RB certification. It is about time and thanks to the Training Agencies for their new expertise and important service [13, 14].

Dive Computer Schematic

Decompression computers are moderately expensive items these days. Basically a decompression meter is a microprocessor computer consisting of a power source, pressure transducer, analog to digital signal converter, internal clock, microprocessor chip with RAM (random access memory) plus ROM (read only memory) and pixel display screen. Pressure readings from the transducer are converted to digital format by the converter and sent to memory with the elapsed clock time for model calculations, usually every $5–10\,sec$. Results are displayed on the screen including time remaining, time at a stop, tissue gas buildup, time to fly and other flag points usually tissue variables, ascent rates, stops and warnings to name a few. Some $3–9\,volts$ is sufficient power to drive the computer for a couple of years assuming about 100 dives per year. The ROM contains the model program (step application of model equations), all constants and queries, the transducer and clock. The RAM maintains storage registers for all dive calculations ultimately sent to the display screen. Dive computers can be worn on the wrist, incorporated in consoles or even integrated into $heads - up$ displays in masks. A typical dive computer is schematized in Fig. 1. Simplification is obvious.

Diving Protocols

Within algorithm implementations and related practices, dive computers operate in modes consistent with protocols for safe and sane diving [2, 4, 9, 10]:

- reduced nonstop time limits consistent with Doppler bubble measurements
- safety stops in the shallow zone for $1–2\,min$ for all diver ascents
- deep 1/2 stops for $1–4\,min$ even on top of scheduled decompression staging
- exploding usage of nitrox and enriched breathing mixtures in recreational diving
- safe altitude diving extensions of sea level protocols
- in recreational diving, computers have supplanted dive tables
- technical divers are relying more and more on dive computers for planning and staging instead of wrist slates

Fig. 1 Dive computer
schematic

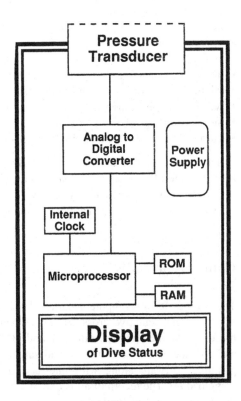

- deep switches to nitrogen based breathing mixtures are avoided by technical divers with a better strategy of increasing oxygen fraction with commensurate decrease in helium fraction in the breathing mixture (helium-oxygen mirroring)
- decompression divers on OC switch to EAN80 or pure oxygen in the 20–30 fsw zone
- RB usage is increasing across the full spectrum of diving
- wrist dive computers possess chip speeds that allow full resolution and implementation of the most complex diving algorithms
- trimix is the gas of choice for deep and decompression diving
- reports suggest that helium mixtures in decompression diving leave the diver feeling better than nitrogen mixtures;
- the computer industry at large is becoming increasingly interested in marketing new dive computers

Powerful and useful as dive computers may be, there are some downsides with their reported usage:

- pushing the computer beyond its model limits and correlation envelope
- not reading the operating manual
- ignoring warnings
- violating ascent rates

- diving a computer that is not properly initialized
- ignoring ceilings
- using one computer for two divers
- improperly entering gas mixtures and pp_{O_2} setpoints
- dialing very liberal user options and factors
- violating depth restrictions
- not performing predive planning
- turning the unit off when in ERROR mode (less possible these days)

Dive computers of the future will build on present generation of models and algorithms to offer divers the most advanced hardware and software technology for the safest and most efficient ways to dive.

A compilation of useful and popular modern dive computers is up next.

Commercial Units

The number of dive computers marketed has grown significantly in the past 20 years or so. A representative cross section of commercial units presently marketed is listed. Units incorporate both GM and BM protocols. These units are modern and engineered for performance and safety. Most have PC connectivity and dive planning software along with interfaces to DAN and LANL DBs for profile downloading. The record of all is one of safe and extensive real world diving under many environmental conditions:

1. **Suunto** Suunto markets a variety of computers all using the RGBM. The EON Steel and DX can be used in gauge, air, nitrox, trimix, OC and CCR modes. The D6, D4 and Vyper are OC computers in gauge, air and nitrox modes. The Zoop and Cobra are recreational computers for gauge, air and nitrox use.
2. **Mares** All Mares computers use the RGBM. Recreational models include the ICON HD, Matrix, Smart and Puck Pro for OC in gauge, air and nitrox modes.
3. **Uwatec** Uwatec computers are marketed by Scuba Pro and all use the ZHL algorithm. The M2 and Pro Mantis are targeted for both recreational and technical diving with gauge, air, nitrox, trimix and CCR modes. The Pro Galileo Sol is a technical dive computer with gauge, air, nitrox and trimix capabilities.
4. **UTC** UTC markets a message sending-receiving computer called the UDI for air and nitrox. All UDIs employ the RGBM. The message exchanging capabilities extend out to $2\,miles$ using sonar, GPS and underwater communications systems. Models include the UDI 14 and UDI 28. Underwater special military units, search and recovery teams and exploration operations use the UDIs routinely today. UDIs also have high resolution compasses for extended navigation.
5. **Huish/Atomic Aquatics/Liquivision** Huish Outdoors owns both Atomic Aquatics and Liquivision. Atomic Aquatics markets a recreational dive computer using the RGBM called the Cobalt for air and nitrox. Liquivision

models include the Kaon, Lynx, X1 and Xeo. The Lynx and Kaon are technical and recreational computers for gauge, air and nitrox modes using the ZHL with GFs. The X1 and Xeo are full up technical dive computers for air, nitrox, trimix and CCR using offering both the ZHL with GFs and RGBM.

6. **Cressisub** All Cressisub computers use the RGBM in recreational gauge, air and nitrox modes. The Newton Titanium. Goa, Giotta and Leonardo are Cressisub models. Cressisub markets a complete line of diving gear in addition to dive computers.

7. **Sherwood** Sherwood computers all use the ZHL. Recreational models for air and nitrox include the Amphos and Wisdom computers.

8. **Oceanic** Oceanic computers use the DSAT and ZHL algorithms for recreational diving. Many models are marketed for gauge, air and nitrox diving and include the VTX, Datamax, Geo, Pro Plus. OCi, Atom, Veo and F10.

9. **Shearwater** Shearwater computers are targeted for technical diving. All use the ZHL with GFs and VPM may be downloaded as an option, The Petrel, Perdex and Nerd2 models address air, nirox, trimix and CCR. Some RB Manufacturers are integrating Shearwater computers into their RB units.

10. **Ratio** Ratio computers employ the ZHL and VPM algorithms for technical and recreational diving. Models include the iX3M Pro and IX3M GPS (Easy, Deep, Tech+, Reb versions) plus the iDive Sport and iDive Avantgarde (Easy, Deep, Tech+ versions) series with air, nitrox, helium and CCR capabilities and GPS and wireless connectivity. The model list is impressive and complete with a strong offering of technical and professional diving units.

11. **Cochrane** Cochrane computers are marketed for recreational and technical diving using the USN LEM (VVAL18). The EMC16 a is recreational air and nitrox computer. The EMC20H is a technical air, nitrox and helium unit. Military units include the EODIII for USN EOD operations and the NSWIII for USN Special Warfare (SEAL) evolutions.

12. **Aeris** Aeris computers are directed at recreational divers using (modified USN) DSAT algorithms for air and nitrox. Models include the A100, A300, A300AI, XR1, NXXR2, Elite T3, Epic and Manta.

Most dive computers are manufactured by one of 4 companies, namely Seiko, Timex, Citizen and Casio, certainly a storied and well known group of fine instrument makers to be sure.

References

1. Brubakk AO and Neuman TS, *Physiology and Medicine of Diving*, Saunders Publishing, 2003, London.
2. Hills BA, *Decompression Sickness: The Biophysical Basis of Prevention and Treatment*, Wiley and Sons Publishing, 1977, Bath.
3. Bove AA and Davis JC, *Diving Medicine*, Saunders Publishing, 2004, Philadelphia.
4. Wienke BR, *Science of Diving*, CRC Press, 2015, Boca Raton.

5. Joiner JJ, *NOAA Diving Manual: Diving for Science and Technology*, Best Publishing, 2001, Flagstaff.
6. Wienke BR, *Biophysics and Diving Decompression Phenomenology*, Bentham Science Publishers, 2016, Sharjah.
7. Schreiner HR and Hamilton RW, *Validation of Decompression Tables*, Undersea and Hyperbaric Medical Society Workshop, 1989, UHMS Publication 74(VAL) 1-1-88, Washington DC.
8. Wienke BR, *Basic Decompression Theory and Application*, Best Publishing, 2003, San Pedro.
9. Wienke BR and Graver DL, *High Altitude Diving*, NAUI Technical Publication, 1991, Montclair.
10. Blogg SL, Lang MA and Mollerlokken A, *Validation of Dive Computers Workshop*, 2011, EUBS/NTNU Proceedings, Gdansk.
11. Lang MA and Hamilton RW, *AAUS Dive Computer Workshop*, University of Southern California Sea Grant Publication, 1989, USCSG-TR-01-89; Los Angeles.
12. Vann RD, Dovenbarger J and Wachholz, *Decompression Sickness in Dive Computer and Table Use*, DAN Newsletter 1989; 3-6.
13. Westerfield RD, Bennett PB, Mebane Y and Orr D, *Dive Computer Safety*. Alert Diver 1994; Mar-Apr: 1-47.
14. Bennett PB, Wienke BR and Mitchell S, *Decompression and the Deep Stop Workshop*, 2008, UHMS/NAVSEA Proceedings, Salt Lake City.

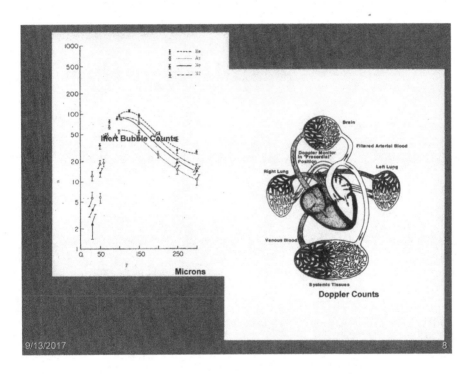

Inert Bubble Counts

Microns

Doppler Counts

Basic Computer Models

Instantaneous estimates of parameters needed to stage divers by dive computers rely an mathematical relationships coupled to pressure sensors and clocks in the unit. The two basic ones, GM and BM, follow [1, 4–6, 8–10] as well as quantification of oxygen toxicity in diving [2, 3, 6, 12, 13].

Dissolved Gas Model (GM)

Diver staging in the classical Haldane approach limits inert gas tissue tensions, p, across all tissue compartments, λ, with halftimes, τ, by a limit point, called the critical tension, M, according to,

$$p = p_a + (p_i - p_a) \, exp \, (-\lambda t) \leq M$$

with p_a ambient gas partial pressure, p_i initial partial tension, t exposure time at p_a and,

$$\lambda = \frac{0.693}{\tau}$$

Equivalently, in terms of tissue gradients, g,

$$g = g_i \, exp \, (-\lambda t) \leq G$$

for,

$$g = p - p_a$$

$$g_i = p_i - p_a$$

© The Author(s), under exclusive licence to Springer Nature Switzerland AG 2018

B. R. Wienke, T. R. O'Leary, *Understanding Modern Dive Computers and Operation*, SpringerBriefs in Computer Science, https://doi.org/10.1007/978-3-319-94054-0_3

with limiting critical gradient, G,

$$G = M - P$$

Halftimes range,

$$3 \leq \tau \leq 540 \, min$$

in applications and critical tensions, M, are linear functions of depth, d, with roughly across GM models,

$$M = \tau^{-0.25}(153.3 + 4.11d) \, fsw$$

If M is exceeded at any point on ascent, a decompression stop is required. According to fits [4, 6] across popular GM models, an approximate relationship between tissue halftime, τ, and nonstop limit, t_n, can be written,

$$\lambda t_n = 1.25$$

Helium tissue halftimes are 1/3 nitrogen tissue halftimes. GM algorithm are used in recreational and technical diving across OC and RB systems.

The GM algorithm typically brings diver into the shallow zone for decompression (shallow stops). Ascent rates are nominally a slow 30 fsw/min. Critical tensions, M, have little to nothing to do with actual bubble formation in the tissue and blood but are (statistical) medical limit points to observed nonstop diving outcomes using basically arbitrary tissue compartments, τ. The approach dates back to Haldane and the 1900s and has been used extensively since then with little change and some tweaking of values. Much testing by world Navies has ensued on the medical side. Notable are the USN [14] and ZHL[15] models and protocols about particulars follow as used in dive computers.

1. **USN Model (Workman [14])** In the Workman USN approach, the permissible gas tension, Π (nitrogen plus helium), is limited by,

$$\Pi \leq M$$

with M critical tensions listed in Table 1 for depth, d,

$$M = M_0 + \Delta M d$$

where depth, d, is the difference between total ambient pressure, P, and surface pressure, P_0,

$$d = P - P_0$$

Table 1 Workman USN
M-Values

Nitrogen			Helium		
τ_{N_2} (min)	M_0 (fsw)	ΔM	τ_{He} (min)	M_0 (fsw)	ΔM
5	104	1.8	5	86	1.5
10	88	1.6	10	74	1.4
20	72	1.5	20	66	1.3
40	56	1.4	40	60	1.2
80	54	1.3	80	56	1.2
120	52	1.2	120	54	1.2
160	51	1.1	160	54	1.1
200	51	1.1	200	53	1.0
240	50	1.1	240	53	1.0

Corresponding permissible gradients, G, then satisfy,

$$G = \Pi - P \leq M - P = (M_0 - \Delta M P_0) + (\Delta M - 1)P$$

with P_0 ambient pressure at the surface as noted,

$$P_0 = 33 \, exp \, (-0.038h)$$

for elevation, h, in multiples of 1000 ft. The correlation parameter with the LANL DB, χ, is,

$$\chi = G - (M - P)$$

with surfacing hit criteria,

$$\Pi > M_0$$

2. **ZHL Model (Buhlmann [15])** The Buhlmann ZHL approach was tested at low altitude and is similar to the Workman USN approach, that is, the permissible gas tension, Π (nitrogen plus helium again), is limited by,

$$\Pi \leq Z$$

with critical tensions, Z, given by,

$$Z = a + \frac{P}{b} = a + \frac{P_0 + d}{b} = Z_0 + \Delta Z d$$

with,

$$Z_0 = a + \frac{P_0}{b}$$

$$\Delta Z = \frac{1}{b}$$

Accordingly, we have,

$$G = \Pi - P \leq a + \left[\frac{1}{b} - 1\right](P_0 + d)$$

for constants, a and b, defining Z at sea level ($P_0 = 33\ fsw$) in Table 2. The expressions put the ZHL Z-value model in the same computational framework as the USN M-value model

The LANL DB correlation parameter, χ, is,

$$\chi = G - (Z - P)$$

with hit criteria at the surface,

$$\Pi > a + \frac{P_0}{b}$$

Table 2 Buhlmann swiss Z-values

Nitrogen			Helium		
τ_{N_2} (min)	$Z_0 = a + 33/b$ (fsw)	$\Delta Z = 1/b$	τ_{He} (min)	$Z_0 = a + 33/b$ (fsw)	$\Delta Z = 1/b$
4.0	106.2	1.91	1.5	134.5	2.36
8.0	83.2	1.54	3.0	102.4	1.74
12.5	73.8	1.39	4.7	89.4	1.53
18.5	66.8	1.28	7.0	79.8	1.38
27.0	62.3	1.23	10.2	73.6	1.32
38.3	58.4	1.19	14.5	68.2	1.25
54.3	55.2	1.15	20.6	63.7	1.21
77.1	52.3	1.12	29.0	59.7	1.17
109.2	49.8	1.09	41.1	57.1	1.14
146.0	48.2	1.08	55.2	55.1	1.12
187.0	46.8	1.07	70.7	54.0	1.11
239.0	45.6	1.06	90.3	53.3	1.10
305.0	44.5	1.05	115.3	53.1	1.09
390.0	43.5	1.04	147.4	52.8	1.09
498.0	42.6	1.04	188.2	52.6	1.08
635.0	41.8	1.03	240.0	52.3	1.07

Bubble Phase Model (BM)

Modern bubble phase models (BM) couple tissue tensions to bubbles directly by assuming an exponential distribution, n, of bubble seeds in radii, r, excited into growth by changing ambient, P and dissolved gas, p, total pressure,

$$n = N \, exp \, (-\zeta r)$$

for N and ζ constants obtained and/or fitted to laboratory or diving data. To date, distributions of bubble seeds have not been measured in vivo. Using the same set of tissue halftimes and inert gas tension equations above in the GM, diver staging in the BM requires the cumulative bubble volume excited into growth by compression-decompression, ϕ, to remain below a critical value, Φ, throughout all points of the dive and in all tissue compartments,

$$\int_t \frac{d\phi}{dt} dt = 4\pi D S \int_t \int_r nr^2 \left[p - P - \frac{2\gamma}{r} \right] dr dt \leq \Phi$$

with D tissue diffusivity, S tissue solubility and γ bubble surface tension. In applications, the critical phase volume, Φ, is near $600 \, microns^3$ and surface tension, γ, is roughly 20 $dyne/cm$. Diffusivity times solubility, DS, is fitted to diving data. If Φ is exceeded at any point on ascent, a decompression stop is necessary. The decompression schedule is computed from permissible tissue-bubble gradients and applied iteratively until the surfacing separated phase volume, ϕ, stays below the limit point, Φ. Convergence of the iterative process usually occurs within 2–3 passes.

BM algorithms are used across recreational and technical diving on both OC and RB systems. Staging starts in the deep zone and continues into the shallow zone (deep stops). Ascent rates are also 30 fsw/min. Bubbles are assumed using realistic properties and exponential distributions in size but have never been really measured in humans. A phase volume limit point, Φ, is extracted from diver exposure profiles in the LANL Data Bank using maximum likelihood (ML) correlation techniques. Gel experiments also deduce a phase volume limit point. Testing is nowhere near as extensive as dissolved gas approaches but is growing. BM models rely on correlations with actual mixed gas profiles across OC and RB, deep and decompression diving on arbitrary breathing mixtures. Application and use is growing, particularly in the technical diving sector over the past 20–25 years with new computers implementing bubble models. In particular, the VPM [16] and RGBM [17] models are noteworthy and used extensively within recreational and technical dive computers as detailed in the following.

An early forerunner of BM algorithms, the Hills thermodynamic model (TM), is also described in for completeness. It is the springboard for later bubble models but has not been encoded into commercial dive computers and software. It also generates deep stops but dive computer encoding is very limited in application.

1. **Varying Permeability Model (Yount [16])** The tissue compartments in the Yount VPM for nitrogen consist of the set,

$$\tau_{N_2} = (1, 2, 5, 10, 20, 40, 80, 120, 160, 240, 320, 400, 480, 560, 720) \; min$$

with the helium compartments scaling,

$$\tau_{He} = \frac{\tau_{N_2}}{3}$$

The VPM model links to bubble experiments in gels and related strata. In gel experiments, Yount divided gas diffusion across bubble interfaces into permeable and impermeable regions. For dive model applications, the regions separate around 165 fsw. Bubbles of nitrogen and helium are excited into growth by pressure changes during the dive from some minimum excitation radius, ϵ, in the 0.5 μm range, with nitrogen bubbles slightly larger than helium bubbles and the excitation radius decreasing with increasing absolute pressure, P. The excitation radius separates growing from shrinking bubbles. The radial bubble distribution, n, in the VPM is given by,

$$n = n_0 \; exp \; (-\beta r)$$

with n_0 an experimental normalization factor for gel sample size and β on the order of $1/\epsilon \; \mu m^{-1}$ for diving applications. The staging protocol in the VPM limits the permissible supersaturation, G to prevent bubble growth on ascent,

$$G = \Pi - P \leq \frac{\gamma}{\gamma_c} \left[\frac{2\gamma_c}{\epsilon} - \frac{2\gamma}{\epsilon_0} \right]$$

with γ the usual bubble surface tension and γ_c the crushing bubble surface tension, roughly 20 and 150 $dyne/cm$ respectively. The radius, ϵ_0, is an experimental metric, somewhere near 0.7 μm. For diving, VPM ascents are limited by G at each stage in the decompression and staging profiles are iterated to convergence across all stops. The LANL DB correlation parameter, χ, in the VPM is,

$$\chi = G - \frac{\gamma}{\gamma_c} \left[\frac{2\gamma_c}{\epsilon} - \frac{2\gamma}{\epsilon_0} \right]$$

and surfacing hit condition,

$$\Pi > P_0 + \frac{\gamma}{\gamma_c} \left[\frac{2\gamma_c}{\epsilon} - \frac{2\gamma}{\epsilon_0} \right]$$

2. **Reduced Gradient Bubble Model (Wienke [17])** Nitrogen tissue compartments in the Wienke RGBM range,

$$\tau_{N_2} = (2, 5, 10, 20, 40, 80, 120, 160, 200, 240, 300) \ min$$

with helium compartments,

$$\tau_{He} = \frac{\tau_{N_2}}{2.65}$$

using the ratio of the square root of atomic weights as the scaling factor. The bubble dynamical protocol in the RGBM model amounts to staging on the seed number averaged, free-dissolved gradient across all tissue compartments, G, for, P, permissible ambient pressure, Π, total inert gas tissue tension, n, excited bubble distribution in radius (exponential), γ, bubble surface tension and, r, bubble radius,

$$G \int_{\epsilon}^{\infty} n dr = (\Pi - P) \int_{\epsilon}^{\infty} n dr \leq \int_{\epsilon}^{\infty} \left[\frac{2\gamma}{r} \right] n dr$$

so that,

$$G = (\Pi - P) \leq \beta \ exp \ (\beta\epsilon) \int_{\epsilon}^{\infty} exp \ (-\beta r) \left[\frac{2\gamma}{r} \right] dr$$

for ϵ the excitation radius at P. Time spent at each stop is iteratively calculated so that the total separated phase, Φ, is maintained at, or below, its limit point. This requires some computing power, but is attainable in diver wrist computers presently marketed with the same said for the VPM. The USN and ZHL models are less complex for computer implementation. The limit point to phase separation, Φ, is near 600 μm^3, and the distribution scaling length, β, is close to 0.60 μm^{-1} for both nitrogen and helium. Both excitation radii, ϵ, and surface tension, γ, are functions of ambient pressure and temperature and not constant. The equation-of-state (EOS) assigned to the bubble surface renders the surface tension below lipid estimates, on the order of 15 $dyne/cm$ and excitation radii are below 1 μm. LANL DB correlation parameter, χ, in the RGBM is given by,

$$\chi = G - \beta \ exp \ (\beta\epsilon) \int_{\epsilon}^{\infty} exp \ (-\beta r) \left[\frac{2\gamma}{r} \right] dr$$

and surfacing hit criteria has,

$$\Pi > P_0 + \beta \ exp \ (\beta\epsilon) \int_{\epsilon}^{\infty} exp \ (-\beta r) \left[\frac{2\gamma}{r} \right] dr$$

3. **Thermodynamic Model (Hills [19])** The thermodynamic model (TM) potentially couples both tissue diffusion and blood perfusion equations though the original Hills formulation with application to diving focused only on diffusion. Cylindrical symmetry is assumed in the model. From a boundary vascular zone of thickness, a, gas diffuses into the extended extravascular region, bounded by b. If we wrap the planar geometry into a hollow cylinder of inner radius, a and outer radius, b, we generate what is called Krogh annular geometry. The hollow cylindrical model retains all the features of a planar model and additionally includes curvature for small a and b, with $l = b - a$. Assigning the same boundary conditions at a and b, namely, the tissue tension, p, equals the arterial tension, p_a, writing the diffusion equation in radial cylindrical coordinates,

$$D\frac{\partial^2 p}{\partial r^2} + \frac{D}{r}\frac{\partial p}{\partial r} = \frac{\partial p}{\partial t}$$

and solving yields,

$$p - p_a = (p_i - p_a)\sum_{n=1}^{\infty} X_n\, U_0(\epsilon_n r)\, exp\,(-\epsilon_n^2 Dt)$$

with X_n a constant satisfying initial conditions, U_0 the cylinder functions, and ϵ_n the eigenvalues satisfying,

$$U_0(\epsilon_n a) = \frac{\partial U_0(\epsilon_n b/2)}{\partial r} = 0$$

Averaging over the tissue region, $a \leq r \leq b$, finally gives,

$$p - p_a = (p_i - p_a)\frac{4}{(b/2)^2 - a^2}\sum_{n=1}^{\infty}\frac{1}{\epsilon_n^2}\frac{J_1^2(\epsilon_n b/2)}{J_0^2(\epsilon_n a) - J_1^2(\epsilon_n b/2)}exp\,(-\epsilon_n^2 Dt)$$

with J_1 and J_0 Bessel functions, order 1 and 0. Typical vascular parameters are bounded roughly by,

$$0 < a \leq 4\,\mu m$$

$$10 \leq b \leq 32\,\mu m.$$

Perfusion limiting is applied as a boundary condition through the venous tension, p_v, by enforcing a mass balance across both the vascular and cellular regions at a, according to Hills,

$$\frac{\partial p_v}{\partial t} = -\kappa(p_v - p_a) - \frac{3}{a}S_p D\left[\frac{\partial p}{\partial r}\right]_{r=a}$$

with S_p the ratio of cellular to blood gas solubilities, κ the perfusion constant and p_a the arterial tension. The coupled set relate tension, gas flow, diffusion and perfusion and solubility in a complex feedback loop. The thermodynamic trigger point for decompression sickness is the volume fraction, χ, of separated gas, coupled to mass balance. Denoting the separated gas partial pressure, P_{N_2}, under worse case conditions of zero gas elimination upon decompression, the separated gas fraction is estimated,

$$\chi P_{N_2} = S_c(p - P_{N_2})$$

with S_c the cellular gas solubility. The separated nitrogen partial pressure, P_{N_2} is taken up by the inherent unsaturation and given by (fsw),

$$P_{N_2} = P + 3.21$$

in the original Hills formulation but other estimates have been employed. Mechanical fluid injection pain, depending on the injection pressure, δ, can be related to the separated gas fraction, χ, through the tissue modulus, K,

$$K\chi = \delta$$

so that a decompression criteria requires the staging paradigm,

$$K\chi \leq \delta$$

with δ in the range, for $K = 3.7 \times 10^4 \, dyne \, cm^{-2}$,

$$0.34 \leq \delta \leq 1.13 \, fsw.$$

The popular USN, ZHL, VPM and RGBM algorithms [19, 21–35] implemented across the majority of marketed and tested dive computers have seen widespread and safe usage over many years with GM computers around since the 1970s and BM computers more recent and gaining in popularity since the 1990s especially in the technical diving community. Correlations and tests for USN and ZHL staging are referenced [7, 14, 15] and discussed elsewhere while correlations and tests for VPM and RGBM staging are similarly referenced [4, 18, 20] and discussed in the same frameworks. With extensive computer implementation and safe utilization without noted DCS nor oxtox spikes and staging issues, they all can be considered user validated across nominal recreational and technical diving. It is also reasonable to assume they can and will be safely modified to accommodate diving *beyond the envelope* in the future.

Oxygen Toxicity (OT)

Both pulmonary and CNS toxicity are tracked by dive computers in a relatively simple way [4, 6]. Pulmonary toxicity is tracked with a dose-time estimator, Γ, written,

$$\Gamma = \sum_n \left[\frac{pp_{O_2} - 0.5}{0.5} \right]_n^{0.83} t_n \leq 750 \ min$$

with, pp_{O_2}, oxygen tension (atm) and, t, exposure time (min). Dive segments, n, are tallied every 5–10 sec and Γ updated. Central nervous system toxicity is similarly tallied over dive segments, n, by a CNS clock, Ω, using the oxygen limit points, t_{O_2}, for exposure to oxygen partial pressure, pp_{O_2} (atm), for time, t (min),

$$\Omega = \sum_n \left[\frac{t}{t_{O_2}} \right]_n \leq 1$$

with approximate CNS oxygen time limits (min),

$$t_{O_2} = 4140 \ exp \ (-2.7 pp_{O_2}) \ min$$

In both cases, violations of OT limit points result in dive computer warnings. Variations in tested oxygen limit points are greater than variations in tested nonstop limits in air and nitrox exposures. This is probably a reason why technical divers often exceed CNS oxtox clock limits by large amounts, that is, in the 2–3 range from analysis of some reports. With further testing in the future, one might reasonably expect some tuning of the oxtox dose-time relationships. Without noted oxygen toxicity issues in computer users of the above oxygen dose-time relationships, the present oxtox model seems safe and user validated across popular GM and BM computers on the whole. Reports of oxygen toxicity in divers are fewer and far more between than DCS [3, 11].

But there is certainly more to the story as far as table, software and computer implementations. To encompass far reaching and often diverse changes in diving activities within a unified framework requires more than the simple GM, BM and oxtox models discussed. To track gas dynamics modelers and table designers need address both free and dissolved gas phases, their interplay and their impact on diving protocols. They can't do all so ad hoc procedures are prudent and necessary for safe and efficient diving and ascent staging. Biophysical models of inert gas transport and bubble formation all try to prevent decompression sickness for sure but many snags remain. Developed over years of diving application, biophysical models differ on a number of basic issues still mostly unresolved today:

- the rate limiting process for inert gas exchange, blood flow rate (perfusion) or gas transfer rate across tissue (diffusion)

- composition and location of critical tissues (bends sites)
- the mechanistics of phase inception and separation (bubble formation and growth)
- the critical trigger point best delimiting the onset of symptoms (dissolved gas buildup in tissues, volume of separated gas, number of bubbles per unit tissue volume, bubble growth rate to name a few)
- the nature of the critical insult causing bends (nerve deformation, arterial blockage or occlusion, blood chemistry or density changes or all of the above)

Such issues confront every modeler and table designer, perplexing and ambiguous in their correlations with experiment and nagging in their persistence. And here comments are confined just to Type I (limb) and II (CNS) bends to say nothing of other complications and factors. These concerns translate into a number of what decompression modelers call dilemmas that limit or qualify their best efforts to describe decompression phenomena. Ultimately, such concerns work their way into dive computers, tables and software with the same caveats. Even if mechanisms are better understood, implementations are often too complex computationally for inclusion in meaningful quantitative syntheses. Some of the ad hoc prescriptions that follow fall into that category. Dive computers and diveware for GM, BM and OT computational paradigms offer optimal platforms for safe diving practice. Tables are limited certainly by comparison.

References

1. Brubakk AO and Neuman TS, *Physiology and Medicine of Diving*, Saunders Publishing, 2003, London.
2. Hills BA, *Decompression Sickness: The Biophysical Basis of Prevention and Treatment*, Wiley and Sons Publishing, 1977, Bath.
3. Bove AA and Davis JC, *Diving Medicine*, Saunders Publishing, 2004, Philadelphia.
4. Wienke BR, *Science of Diving*, CRC Press, 2015, Boca Raton.
5. Joiner JJ, *NOAA Diving Manual: Diving for Science and Technology*, Best Publishing, 2001, Flagstaff.
6. Wienke BR, *Biophysics and Diving Decompression Phenomenology*, Bentham Science Publishers, 2016, Sharjah.
7. Schreiner HR and Hamilton RW, *Validation of Decompression Tables*, Undersea and Hyperbaric Medical Society Workshop, 1989, UHMS Publication 74(VAL) 1-1-88, Washington DC.
8. Wienke BR, *Basic Decompression Theory and Application*, Best Publishing, 2003, San Pedro.
9. Wienke BR and Graver DL, *High Altitude Diving*, NAUI Technical Publication, 1991, Montclair.
10. Blogg SL, Lang MA and Mollerlokken A, *Validation of Dive Computers Workshop*, 2011, EUBS/NTNU Proceedings, Gdansk.
11. Lang MA and Hamilton RW, *AAUS Dive Computer Workshop*, University of Southern California Sea Grant Publication, 1989, USCSG-TR-01-89; Los Angeles.
12. Vann RD, Dovenbarger J and Wachholz, *Decompression Sickness in Dive Computer and Table Use*, DAN Newsletter 1989; 3-6.
13. Westerfield RD, Bennett PB, Mebane Y and Orr D, *Dive Computer Safety*. Alert Diver 1994; Mar-Apr: 1-47.

14. Workman RD, *Calculation of Decompression Schedules for Nitrogen-Oxygen and Helium-Oxygen Dives*, 1965, USN Experimental Diving Unit Report, NEDU, 6–65, Washington DC.

15. Buhlmann AA, *Decompression: Decompression Sickness*, Springer-Verlag Publishing, 1984, Berlin.

16. Yount DE and Hoffman DC, *On the Use of a Bubble Formation Model to Calculate Diving Tables*, Aviat. Space Environ. Med. 1986; 36: 149–156.

17. Wienke BR, *Reduced Gradient Bubble Model in Depth*, CRC Press, 2003, Boca Raton.

18. Wienke BR. *Computer Validation and Statistical Correlations of a Modern Decompression Diving Algorithm*. Comp. Biol. Med. 2010; 40: 252–270.

19. Wienke BR and O'Leary TR, *Diving Decompression Models and Bubble Metrics: Dive Computer Syntheses*. Comp. Biol. Med. 2009; 39: 309–311.

20. Wienke BR, *Deep Stop Model Correlations*, J. Bioeng. Biomed. Sci. 2015; 5: 12–18.

21. Kahaner D, Moler C and Nash S, *Numerical Methods and Software: CLAMS*, Prentice-Hall, 1989, Engelwood Cliffs.

22. Le Messurier DH and Hills BA, *Decompression Sickness: A Study of Diving Techniques in the Torres Strait*, Hvaldrat. Skrif. 1965; 48, 54–84.

23. Strauss RH, *Bubble Formation In Gelatin: Implications for Prevention Of Decompression Sickness*, Undersea Biomed. Res. 1974; 1, 169–174.

24. Farm FP, Hayashi EM and Beckman EL, *Diving and Decompression Sickness Treatment Practices Among Hawaii's Diving Fisherman*, University of Hawaii Sea Grant Report, 1986, UNIHI-SEAGRANT-TP-86–01, Honolulu.

25. Kunkle TD and Beckman EL, *Bubble Dissolution Physics and the Treatment of Decompression Sickness*, Med. Phys. 1983; 10, 184–190.

26. Krasberg A, *Saturation Diving Techniques*, Proceedings Fourth International Congress Biometrology, Rutgers University Press, 1966, New Brunswick.

27. Bennett PB, Marroni et al, *Effect of Varying Deep Stop Times and Shallow Stop Times on Precordial Bubble Scores After Dives to 35 msw*, Undersea Hyper. Med. 2007; 31, 399–406.

28. Marroni A, Bennett PB et al *A Deep Stop During Decompression from 82 fsw Significantly Reduces Bubbles and Fast Tissue Tensions*, Undersea Hyper. Med. 2004; 31, 233–243.

29. O'Leary TR, *NAUI Technical Diving Manual*, 2011, NAUI Worldwide Publication, Tampa.

30. Brubakk AO, Amtzen AJ, Wienke BR and Koteng S, *Decompression Profile and Bubble Formation after Dives with Surface Decompression: Experimental Support for a Dual Phase Model of Decompression*, Undersea Hyper. Med. 2003; 30, 181–193.

31. Bennett PB and Vann RD, *Workshop on Decompression Procedures for Depths in Excess of 400 fsw*, 1975, UHMS Publication 98 (VAL), Washington DC.

32. Balestra C, *Validation of Dive Computers Workshop*, 2010, DAN-DSL Proceedings, Gdansk.

33. Thalmann ED, *Phase II Testing of Decompression Algorithms for Use in the US Navy Underwater Decompression Meter*, 1984, NEDU Report 1–84, Panama City.

34. Doolette DJ, Gerth WA and Gault KA, *Redistribution of Decompression Stop Time from Shallow to Deep Stops Increases Incidence of Decompression Sickness in Air Decompression Dives*, 2011, NEDU Report 2011–06, Panama City.

35. Bennett PB, Wienke BR and Mitchell S, *Decompression and the Deep Stop Workshop*, 2008, UHMS/NAVSEA Proceedings, Salt Lake City.

Ad Hoc Dive Computer Protocols

Much like functional hook or crook approaches of Australian pearl divers [2], ad hoc diving protocols on top of existing procedures have surfaced in the last 20 years or so. To date, none of the ad hoc protocols have been tested nor correlated with actual data. Anecdotally, these procedures seem to work though, at least reports contraindicating their usage are not generally recorded. That is certainly good. Just briefly, some important ones are mentioned. The technical diving community has been the primary pusher of many of these protocols. In varying degrees they have been embedded into both GM and BM dive computers.

Best Diving Mixtures

Over the years, technical divers and training have proposed operating rules for limiting both oxygen partial pressure, pp_{O_2}, and nitrogen partial pressure, pp_{N_2}, in diving mixed gases like nitrox, trimix and heliox. The procedures are well known and incorporated into GM and BM dive computers.

The procedure is straightforward across commercial, military and technical diving sectors and goes like this, neglecting water vapor, carbon dioxide and all other trace gases for ambient surface pressure, P_h, and depth, d, measured in fsw and gas partial pressures, pp_{O_2}, pp_{N_2} and pp_{He} given in atm:

- determine oxygen fraction, f_{O_2}, by specifying the maximum partial pressure, pp_{O_2}, supported by the bottom depth and duration of the dive

$$f_{O_2} = \frac{pp_{O_2}}{1 + d/33}$$

with pp_{O_2} in the 0.5–1.4 atm range

B. R. Wienke, T. R. O'Leary, *Understanding Modern Dive Computers and Operation*, SpringerBriefs in Computer Science, https://doi.org/10.1007/978-3-319-94054-0_4

- determine nitrogen fraction, f_{N_2}, by specifying maximum partial pressure, pp_{N_2}, below narcosis threshold,

$$f_{N_2} = \frac{pp_{N_2}}{1 + d/33}$$

with pp_{N_2} somewhere in the 3.0–5.5 atm range
- determine helium fraction, f_{He}, by subtracting oxygen and nitrogen fraction from 1,

$$f_{He} = 1 - f_{O_2} - f_{N_2}$$

and is the expensive part of the dive mixture

The same procedure is applied to gas switches and setpoint switches on the way up (ascent) permitting decompression and oxygen management across the whole profile, top to bottom and then bottom to top. As such, it is an essential ingredient for decompression and extended range dive planning on mixed gases. And it applies to OC and RB diving. Some dive computers and coupled dive planning software incorporate best diving mixture strategies.

Altitude Modifications

Altitude modifications are not really ad hoc but are based on simple altitude-pressure relationships [9, 15] and proper extrapolations of critical parameters to reduced pressure. The extrapolations are different for GM and BM computers but based on ambient pressure, P, and elevation, h, according to the barometer equation,

$$P = P_0 \, exp \, (-0.038h)$$

for sea level ambient pressure, $P_0 = 33 \, fsw$, and elevation, h, measured in multiples of 1,000 ft. The exponential term, α,

$$\alpha = \frac{P}{P_0} = exp \, (-0.038h)$$

plays the role of scaling factor for both GM and BM algorithms.

For M-value models, the scaling affects the surfacing M-value, M_0, only and we have,

$$M = \alpha M_0 + \Delta M d$$

with generically,

$$M_0 = \tau^{-0.25} 153.3$$

$$\Delta Md = \tau^{-0.25} 4.11d$$

for depth, d, M- value change per unit depth, Δ, and tissue halftime, τ. Same applies to Z_0 and ΔZ.

Linear extrapolation of M-values and Z-values to altitude (and not the proper exponential decrease of M_0) are only good up to about 15,000 ft and become more incorrect with decreasing ambient pressure. Recall,

$$\lim_{h \to \infty} \frac{P}{P_0} = \lim_{h \to \infty} exp\left(-0.00338h\right) \to 0$$

and all surfacing M-values and Z-values must approach zero as ambient pressure, P, approaches zero. At zero ambient pressure, P, allowable supersaturation must also approaches zero by simple thermodynamic (entropy) constraints (Second Law of Thermodynamics). Linear extensions of M-values and Z-values to altitude as published in the literature are patently wrong on first principles. Simplest fix is to multiply M_0 and Z_0 by $exp\left(-0.00338h\right)$.

For BM algorithms, the extrapolation merely amounts to increasing sea level bubble excitation radii, ϵ_0, by α^{-1},

$$\epsilon = \alpha^{-1} \epsilon_0$$

rendering bubbles larger and decompression times longer. The reduction in pressure also impacts the material properties of bubbles but rather slightly. Boyle expansion of bubbles on ascent is built into some BM computers but if not, Boyle expansion can be tracked at gas switches or other select points. Bubbles are not ideal gases and an appropriate EOS must be employed to tag response of bubble surfaces and volumes under pressure changes. Gas diffusivities, D across bubble interfaces change slightly with elevation as do surface tensions, $2\gamma/r$, for bubble radii of curvature, r. In present BM algorithms, the permissible separated phase, Φ, remains unchanged at altitude.

Pyle 1/2 Stops

Pyle is a diving specimen fisherman out of the Bishop Laboratories at the University of Hawaii. Pyle pioneered the technique of making 1/2 stops for minutes on top of decompression requirements within Haldane dissolved gas tables [1, 6]. Stop times vary from a few minutes at the deepest stop to minutes on the way up the 1/2 stop ladder to the surface. Some of the profiles with 1/2 stops mimic the VPM and RGBM in broadest features. The suggested Bennett and Maronni [4] 1/2 deep stop Doppler score protocols were also correlated within BM algorithms and associated tables.

Reduction Factors

Reduction factors (RF) are a published set [8, 17] of M-value multipliers to mimic bubble models. RFs are always less than 1 to restrict reverse profiles, surface intervals less than $60\,min$, ascents faster than $30\,fsw/min$ and deep depth-differing multiday diving. The simple approach defines a new M-value (or Z-value) by multiplying the critical gradient, G, by a scaling factor, χ, so that

$$M = \chi G + P$$

for ambient pressure, P. They are time dependent and coalesce to 1 after 30–$60\,min$ SIs. Specifically, in terms of the short surface interval factor, χ_1, deeper than first dive factor, χ_2, and multiday factor, χ_3, we write,

$$\chi = \chi_1 + \chi_2 + \chi_3 \leq 1$$

with,

$$\chi_1 = 1 - 0.45\,exp\left[-\frac{(t_{SI} - 30)^2}{900}\right]$$

$$\chi_2 = 1 - 0.45\left[1 - exp\left(-\frac{\delta}{P}\right)\right]exp\left[-\frac{(t_{SI} - 60)^2}{3600}\right]$$

$$\chi_3 = 0.70 + 0.30\,exp\left(-\frac{n}{12}\right)$$

which are pseudo-bubble factors applied to GM models. These forms for multidiving, χ, are dependent on time between dives (surface interval), t_{SI}, ambient pressure difference on reverse profile dives, δ, ambient pressure, P, and multiday diving frequency, n, over $24\,hr$ time spans with t_{SI} measured in min and n the number of consecutive days of diving within $24\,hr$ time spans. Factors are applied after $3\,min$ of surface interval. The reverse profile difference, δ, is the time averaged difference between depths on the present and previous dives and is computed on-the-fly. They can be applied to any GM algorithm and computer.

Reduction factors are consistent (folded in maximum likelihood) with the following:

- Doppler bubble scores peak in $30\,min$ or so after a dive
- reverse profiles with depth increments beyond $50\,fsw$ incur increasing DCS risk, somewhere between 5% and 8% in the depth increment range of 40–$120\,fsw$
- Doppler bubble counts drop tenfold when ascent rates drop from 60 to $30\,fsw/min$
- random multiday diving risks increase by factors of 2–3 (though still small) over risk associated with a single dive according to some reports and data

Gradient Factors

Gradient factors (GF) are a spinoff of above published reduction factors (RF) for technical diving [3, 5, 7, 10]. The GFs can reduce dissolved gas limiters at depth thus producing deep stops on top of dissolved gas staging or increase dissolved limiters to give the diver more bottom time and less decompression stop time. They are often employed by technical divers for deep and decompression diving. They can be constructed in principle to mimic BM algorithms, an academic exercise that might be pursued in the future as a safety exercise. GFs less than 1 produce deep stops while GFs greater than 1 give more bottom and less decompression time. They haven't been tested formally but their use seems to be supported by word of mouth and Internet postings. As with RFs, we have

$$M = \xi G + P$$

Usage varies but roughly in some broad range, GFs have,

$$0.60 \leq \xi \leq 1.20$$

and are not time dependent. In technical diving, they are often applied at OC gas switches or CCR setpoint changes on the way up. A user interface permits insertion of ξ in computers and software.

Dynamical Bubble Factors

In BM computers and software, bubble parameters can also be changed for diving. The excitation bubble radius, ϵ, permissible surfacing separated phase, Φ, and Boyle EOS (expansion) properties can be changed any dive. Increases in the excitation radius, ϵ, induces greater stop time and deeper decompression staging. Reduction in the permissible separated phase, Φ, increases stage times. Relaxation of bubble material properties requires more decompression time in the shallow zone due to increased Boyle expansion. These factors are not time dependent like the RFs.

Isobaric Counterdiffusion (ICD) Regimens

As also seen in Table 7 of chapter "Risk Estimators", some ICD prescriptions have surfaced [10, 12–14, 16]. Apart from purposes of illustration, holding the nitrogen fraction constant helps minimize ICD problems on OC with switches according to simple gas transport considerations. There are no large (ingassing) countercurrents of nitrogen on top of outgassing helium, possibly leading to supersaturation and isobaric counterdiffusion problems with bubbles. On RB systems, of course, there are no real ICD problems on fixed diluents. Suggested is a hierarchal scheme:

- zero rule – no N_2 switches
- first rule – no N_2 switches below 100 fsw
- second rule – no N_2 switches below 70 fsw and END less than 50 fsw

The scheme developed in the field tries to minimize or eliminate the isobaric supersaturation condition depicted in Fig. 1. These have not been formally tested nor linked to helium-oxygen mirroring. From just a gas transport perspective, mirroring is an optimal staging procedure as can be seen with some dive computers or most dive planning software packages.

Helium-Oxygen Mirroring and ICD Mitigation

A practice has developed among seasoned OC divers on trimix and heliox mixtures to reduce the helium fraction, f_{He}, and increase the oxygen fraction, f_{O_2}, in the same proportion with gas mix switches on ascent. An example of this is seen in Table 5 of chapter "Risk Estimators". At the same time, the nitrogen fraction, f_{N_2}. is kept relatively constant to avoid *isobaric slams* as they are popularly

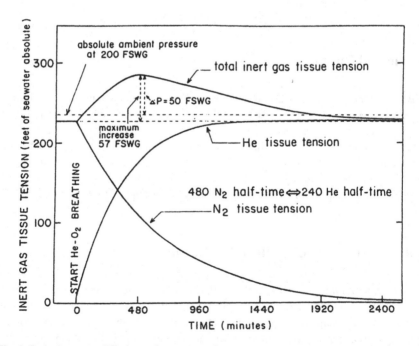

Fig. 1 Isobaric counterdiffusion and supersaturation

termed because of possible large mismatches in gas fractions at switch points. This has not been tested of course but seems a sane approach which also minimizes decompression time. Some GM and BM dive computers include such dive planning in their drop down menus but the practice is not universal nor necessarily mandatory. Coupled to helium-oxygen mirroring might be a switch in the $70\,fsw$ zone to EAN50 and another switch to pure oxygen or EAN80 at the 20–$30\,fsw$ depth. It is the higher concentration of oxygen that helps reduce decompression time and not the isobaric switch to a nitrogen based mixture that is often suggested. If the breathing mixture were heliox (no nitrogen) the best strategy would be EAH50 and EAH80 on the way up. Schedules such as these are important in correlating and validating models against computer downloaded profiles [18–20].

Shallow Safety Stops

Based on suggestions at an America Academy Of Underwater Sciences (AAUS) Workshop [11] discretionary safety stops for 2–$3\,min$ in the 10–$20\,fsw$ zone are recommended for GM and BM staging. Calculations reported [4, 8] underscore the bases of suggestions for a number of reasons. Relative changes in three computed trigger points, tissue tension, separated phase volume and bubble radius for a nominal air bounce dive to $120\,fsw$ for $12\,min$, with and without a safety stop at $15\,fsw$ for $3\,min$, shows marked reductions in bubble and phase volume growth while accommodating insignificant levels of dissolved gas buildup in the slow tissue compartments. The reduction in growth parameters far outstrips any dissolved gas buildup in slow compartments and faster compartments naturally eliminate dissolved gases during the stop, important for deeper diving. Table 1 summarizes calculations and pertinent results.

Surfacing tissue tensions can change quickly with safety stops and slower ascent rates. The faster compartments will respond more quickly than the slower ones and the relative change in tissue tensions will be greater in faster compartments. Similarly, bubbles will grow if conditions favor expansion on ascent. Fast ascents

Table 1 Relative decrease in critical parameters after safety stop

τ	$\Delta\Pi/\Pi$	$\Delta\phi/\phi$	$\Delta r/r$	Surfacing Π	Surfacing Π
Halftime				$120\,fsw/15\,min$	$120\,fsw/12\,min/15\,fsw/3\,min$
(min)				(fsw)	(fsw)
5	0.21	0.34	0.68	101.5	77.0
10	0.11	0.24	0.39	87.5	73.0
20	0.06	0.11	0.24	66.9	59.0
40	0.02	0.08	0.18	49.9	45.7
80	−0.01	−0.03	0.02	39.0	36.9
120	−0.02	−0.04	−0.01	34.9	33.5

(decreased average pressure) can maximize their growth rate while slower ascents and safety stops (increased average pressure) can support their dissolution. Bubble growth and dissolved gas buildup compete as seen in Table 1 contrasting relative changes in tissue tensions, $\Delta\Pi/\Pi$, separated phase volumes, $\Delta\phi/\phi$, and bubble radii, $\Delta r/r$, for an air dive to 120 fsw for 15 min and the same dive to 120 fsw for 12 min with a safety stop at 15 $fsw/3\,min$. Clearly effects are greater on phase triggers than critical tensions. Calculations are typical for bounce exposures in the 40–150 fsw range. If stop time is added to bottom time the exposure exhibits higher tensions as seen in columns 5 and 6, contrasting tensions after the safety stop with the actual dive to 120 $fsw/15\,min$. Big differences occur in the fastest compartments but the procedure is conservative. A small table penalty in the slowest compartment, incurred by adding stop time to bottom time, is offset by free phase reduction after the stop.

The shallow safety stop prescription almost universally accepted and practiced requires 2–3 min in the 10–20 fsw zone before surfacing. The stop in the shallow zone also encourages and teaches buoyancy control especially for neophyte divers. Safety stops in the shallow zone are implemented in virtually all computers (GM and BM) and missed safety stops are followed by computer audible and visual warnings. Tables notate shallow safety stops and some diveware builds shallow safety stops into their user interface and menus.

User Conservancy Knobs

All algorithm and computer settings can be adjusted to user comfort and safety ranges predetermined by the Vendor and varying across computers. Approximate 10–15% reductions or increases in critical dive parameters can be dialed into computers at the start of the dive and maintained throughout the dive. This applies to both GM and BM computers and model parameters. The range of some critical dive parameter, υ, becomes,

$$0.85\upsilon \le \upsilon \le 1.15\upsilon$$

as a choice on drop down menus of dive computers and some software. The restriction in range, υ, prevents misuse and possible safety issues. Reductions or increases of 10–15% in key GM and BM model parameters have noticeable impact on diving and staging regimens. These *user knobs* have not been formally tested nor evaluated but there is some pressure to do so in the same manner as the nominal parameters are correlated and user validated.

Next up is the important topic of computer correlations and validation with downloaded dive computer profiles in the LANL DB. Data Banks such as those at DAN and LANL are part of the technological evolution given to diving science by dive computers. Impacts are monumental and growing.

References

1. Brubakk AO and Neuman TS, *Physiology and Medicine of Diving*, Saunders Publishing, 2003, London.
2. Hills BA, *Decompression Sickness: The Biophysical Basis of Prevention and Treatment*, Wiley and Sons Publishing, 1977, Bath.
3. Bove AA and Davis JC, *Diving Medicine*, Saunders Publishing, 2004, Philadelphia.
4. Wienke BR, *Science of Diving*, CRC Press, 2015, Boca Raton.
5. Joiner JJ, *NOAA Diving Manual: Diving for Science and Technology*, Best Publishing, 2001, Flagstaff.
6. Wienke BR, *Biophysics and Diving Decompression Phenomenology*, Bentham Science Publishers, 2016, Sharjah.
7. Schreiner HR and Hamilton RW, *Validation of Decompression Tables*, Undersea and Hyperbaric Medical Society Workshop, 1989, UHMS Publication 74(VAL) 1-1-88, Washington DC.
8. Wienke BR, *Basic Decompression Theory and Application*, Best Publishing, 2003, San Pedro.
9. Wienke BR and Graver DL, *High Altitude Diving*, NAUI Technical Publication, 1991, Montclair.
10. Blogg SL, Lang MA and Mollerlokken A, *Validation of Dive Computers Workshop*, 2011, EUBS/NTNU Proceedings, Gdansk.
11. Lang MA and Hamilton RW, *AAUS Dive Computer Workshop*, University of Southern California Sea Grant Publication, 1989, USCSG-TR-01–89; Los Angeles.
12. Vann RD, Dovenbarger J and Wachholz, *Decompression Sickness in Dive Computer and Table Use*, DAN Newsletter 1989; 3–6.
13. Westerfield RD, Bennett PB, Mebane Y and Orr D, *Dive Computer Safety*. Alert Diver 1994; Mar-Apr: 1–47.
14. Workman RD, *Calculation of Decompression Schedules for Nitrogen-Oxygen and Helium-Oxygen Dives*, 1965, USN Experimenal Diving Unit Report, NEDU, 6–65, Washington DC.
15. Buhlmann AA, *Decompression: Decompression Sickness*, Springer-Verlag Publishing, 1984, Berlin.
16. Yount DE and Hoffman DC, *On the Use of a Bubble Formation Model to Calculate Diving Tables*, Aviat. Space Environ. Med. 1986; 36: 149–156.
17. Wienke BR, *Reduced Gradient Bubble Model in Depth*, CRC Press, 2003, Boca Raton.
18. Wienke BR. *Computer Validation and Statistical Correlations of a Modern Decompression Diving Algorithm*. Comp. Biol. Med. 2010; 40: 252–270.
19. Wienke BR and O'Leary TR, *Diving Decompression Models and Bubble Metrics: Dive Computer Syntheses*. Comp. Biol. Med. 2009; 39: 309–311.
20. Wienke BR, *Deep Stop Model Correlations*, J. Bioeng. Biomed. Sci. 2015; 5: 12–18.

Computer Profile Data

To validate computer models [1–13, 18–20], diving data is necessary. In the past, data consisted mostly of scattered open ocean and dry chamber tests of specific dive schedules. In such instances, the business of correlating model and diving data was only scratched. Today, profile collection across diving sectors is proceeding more rapidly. Notable are the efforts [1, 4] of Divers Alert Network (DAN) and Los Alamos National Laboratory (LANL). DAN USA is collecting profiles in an effort called Project Dive Exploration (PDE) and DAN Europe has a parallel effort called Diving Safety Laboratory (DSL). The focus has been recreational dive profiles for air and nitrox. The LANL Data Bank collects profiles from technical diving operations on mixed gases for deep and decompression diving on OC and RB systems. Some interesting features of the data have emerged so far from these collections [11–13]:

- profile collection of diver outcomes is an ongoing effort at DAN USA, DAN Europe and LANL and has aided in model tuning using rigorous statistical techniques
- there are no reported spikes in DCS/OT rates for recreational and technical divers using dive computers
- statistics gathered at DAN and LANL suggest that DCS/OT rates are low across recreational and technical diving, but that technical diving is some 10–20 times riskier than recreational diving
- data from meter Manufacturers and Training Agencies, reported as anecdotal at recent Workshops, suggests the DCS/OT incidence rate is on the order of 143/6,000,000 dives (underlying incidence) for computer users
- the underlying incidences in the DAN and LANL profile data are small on the order of 80/187451 (as of 2016) and 28/3459 respectively

Profile collection efforts such as these can enormously benefit divers and diving science. Without downloadable profile data from dive computers, meaningful

© The Author(s), under exclusive licence to Springer Nature Switzerland AG 2018   45
B. R. Wienke, T. R. O'Leary, *Understanding Modern Dive Computers and Operation*,
SpringerBriefs in Computer Science, https://doi.org/10.1007/978-3-319-94054-0_5

algorithm and protocol analysis is very difficult. Profile data banks are important resources for all kinds of diving

In both GM and BM cases, data collection is an ongoing effort. Profile information can be narrowed down to its simplest form, most of it coming from dive computer downloads tagging information across variable time intervals (3–5 sec) which is then processed into a more manageable format for statistical analysis:

- bottom mix/pp_{O_2}, depth and time
- ascent and descent rates
- stage and decompression mix/pp_{O_2}, depths and times
- surface intervals
- time to fly
- diver age, weight, sex and health complications
- outcome rated 1–5 in order of bad to good
- environmental factors (temperature, current, visibility, equipment)

Different DBs will use variations on reported data but the above covers most of the bases. Consider the LANL DB in more detail [14–17]. Risk estimates to follow are based on downloaded profiles and data in the LANL DB.

LANL DB

Some 3569 profiles now reside in the LANL DB. There are 28 cases of DCS in the data file. The underlying DCS incidence rate is, $p = 28/3569 = 0.0078$, below but near 1%. Stored profiles range from 150 fsw down to 840 fsw, with the majority above 350 fsw. All data enters through the Authors, that is, divers, profiles and outcomes are filtered.

The following summary breakdown of DCS hits (bends) updates our earlier reporting and data consists of the following:

- OC deep nitrox reverse profiles – 5 hits (3 DCS I, 2 DCS II)
- OC deep nitrox – 3 hits (2 DCS I, 1 DCS II)
- OC deep trimix reverse profiles – 2 hits (1 DCS II, 1 DCS III)
- OC deep trimix – 4 hits (3 DCS I, 1 DCS III)
- OC deep heliox – 2 hits (2 DCS II)
- RB deep nitrox – 4 hits (2 DCS I, 2 DCS II)
- RB deep trimix – 4 hits (3 DCS I, 1 DCS III)
- RB deep heliox – 4 hits (3 DCS I, 1 DCS II)

DCS I means limb bends, DCS II implies central nervous system (CNS) bends and DCS III denotes inner ear bends (occurring mainly on helium mixtures). Both DCS II and DCS III are fairly serious afflictions while DCS I is less traumatic. Deep nitrox means a range beyond 150 fsw, deep trimix means a range beyond 200 fsw and deep heliox means a range beyond 250 fsw as a rough categorization. The abbreviation OC denotes open circuit while RB denotes rebreather. Reverse profiles are any sequence of dives in which the present dive is deeper than the previous dive. Nitrox means an oxygen enriched nitrogen mixture (including air), trimix denotes a breathing mixture of nitrogen, helium, oxygen and heliox is a breathing mixture of helium and oxygen. None of the trimix nor heliox cases involved oxygen enriched mixtures on OC and RB hits did not involve elevated oxygen partial pressures above 1.4 *atm*. Nitrogen-to-helium (*heavy-to-light*) gas switches occurred in 4 cases violating contemporary ICD (isobaric counterdiffusion) protocols. Isobaric counterdiffusion refers to two inert gases (usually nitrogen and helium) moving in opposite directions in tissues and blood. When summed, total gas tensions (partial pressures) can lead to increased supersaturation and bubble formation probability as seen in Fig. 1 of chapter "Ad Hoc Dive Computer Protocols".

None of the set exhibited pulmonary (full body) nor CNS (central nervous system) oxygen toxicity (*oxtox*). The 28 cases come after the fact, that is diver distress with hyperbaric chamber treatment following distress. Profiles originate with seasoned divers as well as from broader field testing reported to us and coming from divers using wrist slate decompression tables with computer backups. Most profiles reach us directly as computer downloads which we translate to a requisite format. Approximately 93% of all LANL DB entries emanate from computer downloads.

The data is relatively coarse grained making compact statistics difficult. The incidence rate across the whole set is small, on the order of 1% and smaller. Fine graining into depths will be useful in the following but first breakout of data into gas categories (nitrox, heliox, trimix) is repeated as tabulated earlier. Table 1 indicates the breakdown.

Table 1 Profile gas-DCS summary

Mix	Total profiles	DCS hits	Incidence
OC nitrox	459	8	0.0174
RB nitrox	665	4	0.0060
All nitrox	1124	12	0.0107
OC trimix	771	6	0.0078
RB trimix	869	4	0.0046
All trimix	1640	10	0.0061
OC heliox	166	2	0.0120
RB heliox	639	4	0.0063
All heliox	805	6	0.0075
Total	3569	28	0.0078

The DCS hit rate with nitrox is higher but not statistically meaningful across this sparse set. The last entry is all mixes as noted previously. In the above set, there are 49 *marginals*, that is, DCS was not diagnosed but the diver surfaced feeling badly. In such cases, many do not weight the dive as a DCS hit. Others might weight the dive 1/2.

It is also interesting to break mixed gas profiles into 100 *fsw* increments It is obvious that 500 *fsw* or so is the limit statistically to the data set. It is for that reason

Table 2 Profile gas-depth summary

	100 to 200 fsw	200 to 250 fsw	250 to 300 fsw	300 to 350 fsw	350 to 400 fsw	400+ fsw	Total
OC nitrox	358	101					459
RB nitrox	223	311	131				665
OC trimix	30	443	266	26	4	2	771
RB trimix	32	393	321	113	10		869
OC heliox		62	69	30	5		166
RB heliox	52	230	188	142	17	10	639
Total	695	1540	975	311	36	12	3569

Table 3 Profile gas-depth DCS summary

	100 to 200 fsw	200 to 250 fsw	250 to 300 fsw	300 to 350 fsw	350 to 400 fsw	400+ fsw	Total
OC nitrox	5	3					8
RB nitrox		2	2				4
OC trimix		2	2	1		1	6
RB trimix		2	1	1			4
OC heliox			2				2
RB heliox		1	2	1			4
Total	5	10	9	3		1	28

that we limit applications of the LANL algorithm to 540 fsw in dive computers and planning software.

The corresponding DCS hit summary for Table 2 is given in Table 3.

A few observations about the LANL DB ought be passed back to readers:

- deep stop data is intrinsically different from data collected in the past for diving validation in that previous data is mainly based on shallow stop diver staging, a possible bias in dive planning
- deep stop data and shallow stop data yield the same risk estimates for nominal, shallow and nonstop diving because bubble models and dissolved gas models converge in the limit of small phase separation
- if shallow stop data is employed in all cases covered, dissolved gas risk estimates will be usually higher than those computed with BM algorithms in general
- bubble risks estimated herein are higher than risk estimates in other analyses, perhaps a conservative bias
- data entry in the LANL Data Bank is a ongoing process of profile addition
- extended exposure-depth range and mixed gas diving applications are the richest source of information for model tuning

Dive computers are fairly recent developments on the diving scene with no spikes in DCS nor oxtox incidence rates reported within categories of divers employing them. Yet certainly few algorithms have been fully tested, particularly in the deep, decompression and mixed gas diving zones. This is true for both GM and BM algorithms. In time, algorithms will be further analyzed with growing profile data, and protocols validated, modified or discarded. Of course, wet and dry testing are expensive, limited in range and not always viable operationally. In that respect, profile DBs with diver outcomes are enormously important to cover a full spectrum of diving not amenable nor feasible for wet and dry testing. The profile DBs at DAN and LANL are up and running with DAN focused on recreational diving and LANL concerned with technical deep, mixed gas and decompression diving. To say these DBs help fill holes in the testing arena is an understatement. Real diving in actual underwater environments is hard to duplicate in a dry chamber or wet pod.

References

1. Brubakk AO and Neuman TS, *Physiology and Medicine of Diving*, Saunders Publishing, 2003, London.
2. Hills BA, *Decompression Sickness: The Biophysical Basis of Prevention and Treatment*, Wiley and Sons Publishing, 1977, Bath.
3. Bove AA and Davis JC, *Diving Medicine*, Saunders Publishing, 2004, Philadelphia.
4. Wienke BR, *Science of Diving*, CRC Press, 2015, Boca Raton.
5. Joiner JJ, *NOAA Diving Manual: Diving for Science and Technology*, Best Publishing, 2001, Flagstaff.
6. Wienke BR, *Biophysics and Diving Decompression Phenomenology*, Bentham Science Publishers, 2016, Sharjah.
7. Schreiner HR and Hamilton RW, *Validation of Decompression Tables*, Undersea and Hyperbaric Medical Society Workshop, 1989, UHMS Publication 74(VAL) 1-1-88, Washington DC.
8. Wienke BR, *Basic Decompression Theory and Application*, Best Publishing, 2003, San Pedro.
9. Wienke BR and Graver DL, *High Altitude Diving*, NAUI Technical Publication, 1991, Montclair.
10. Blogg SL, Lang MA and Mollerlokken A, *Validation of Dive Computers Workshop*, 2011, EUBS/NTNU Proceedings, Gdansk.
11. Lang MA and Hamilton RW, *AAUS Dive Computer Workshop*, University of Southern California Sea Grant Publication, 1989, USCSG-TR-01-89; Los Angeles.
12. Vann RD, Dovenbarger J and Wachholz, *Decompression Sickness in Dive Computer and Table Use*, DAN Newsletter 1989; 3–6.
13. Westerfield RD, Bennett PB, Mebane Y and Orr D, *Dive Computer Safety*. Alert Diver 1994; Mar-Apr: 1–47.
14. Workman RD, *Calculation of Decompression Schedules for Nitrogen-Oxygen and Helium-Oxygen Dives*, 1965, USN Experimenal Diving Unit Report, NEDU, 6–65, Washington DC.
15. Buhlmann AA, *Decompression: Decompression Sickness*, Springer-Verlag Publishing, 1984, Berlin.
16. Yount DE and Hoffman DC, *On the Use of a Bubble Formation Model to Calculate Diving Tables*, Aviat. Space Environ. Med. 1986; 36: 149–156.
17. Wienke BR, *Reduced Gradient Bubble Model in Depth*, CRC Press, 2003, Boca Raton.
18. Wienke BR. *Computer Validation and Statistical Correlations of a Modern Decompression Diving Algorithm*. Comp. Biol. Med. 2010; 40: 252–270.
19. Wienke BR and O'Leary TR, *Diving Decompression Models and Bubble Metrics: Dive Computer Syntheses*. Comp. Biol. Med. 2009; 39: 309–311.
20. Wienke BR, *Deep Stop Model Correlations*, J. Bioeng. Biomed. Sci. 2015; 5: 12–18.

Wet and Dry Tests and Data

Testing and validation of GM models has been a successful medical exercise for many years dating back to the 1900s and certainly could fill a book. Some are recounted here and more can be found elsewhere [1, 7, 14, 15, 18, 20]. BM models are newer and do not enjoy the testing history of GM models. Apart from computer profile correlations just described, some wet and dry tests have transpired. For BM computers these are important benchmarks. We start at the beginning in first recounting early deep stop testing by Haldane and the deep stop hook-or-crook protocols of Australian and Hawaiian pearl divers and fishermen.

Haldane Deep Stops

Haldane originally found that deep stops were necessary in deeper decompression tests and staging regimens [1, 3, 5, 6] but either abandoned them or could not incorporate them naturally into his GM algorithm on first principles. Too bad some think, he might have saved future generations of divers scheduling controversies and predicated necessary testing. World Navies at the time never tested deep stops either. Deep stops do not emerge in GMs for deco scheduling except when using RFs and GFs. Maybe Haldane didn't go deep enough to see real diving differences and time savings. Deep stops are really a deep protocol. Having said that, nothing detracts from the original research and pioneering medical work of Haldane for sure.

But even before BM algorithms and deep stop protocols emerged, utilitarian diving practices among diving fisherman and pearl gatherers suggested traditional staging was in need of rethinking. And early deco models such as the so-called thermodynamic model (TM) of Hills suggested why and how. Deep stops resurfaced and evolved from cognizance of operational diving practice. Primitive BM algorithms followed and accelerated with introduction and use of dive computers.

© The Author(s), under exclusive licence to Springer Nature Switzerland AG 2018   53
B. R. Wienke, T. R. O'Leary, *Understanding Modern Dive Computers and Operation*,
SpringerBriefs in Computer Science, https://doi.org/10.1007/978-3-319-94054-0_6

Australian Pearl and Hawaiian Diving Fishermen

Pearling fleets operating in the deep tidal waters off Northern Australia employed Okinawan divers who regularly journeyed to depths of 300 fsw for as long as one hour, two times a day, six days per week and ten months out of the year. Driven by economics and not science these divers developed optimized decompression schedules empirically even with the sad loss of 1000s of lives. What a wet test. As reported and analyzed by Le Messurier and Hills [22], deeper decompression stops but shorter decompression times than required by Haldane theory were characteristics of their profiles. Recorders placed on these divers attest to the fact. Such protocols are consistent with minimizing bubble growth and the excitation of nuclei through the application of increased pressure. Even with a high incidence of surfacing decompression sickness following diving, the Australians devised a simple, but very effective, in-water recompression procedure. The stricken diver is taken back down to 30 fsw on oxygen for roughly 30 min in mild cases, or 60 min in severe cases. Increased pressures help to constrict bubbles while breathing pure oxygen maximizes inert gas washout (elimination). Recompression times scale as bubble dissolution experiments in the lab [23] which is quite extraordinary.

Similar schedules and procedures have evolved in Hawaii among diving fishermen according to Farm and Hayashi [24]. Harvesting the oceans for food and profit, Hawaiian divers make between eight and twelve dives a day to depths beyond 350 fsw. Profit incentives induce divers to take risks relative to bottom time in conventional tables. Repetitive dives are usually necessary to net a school of fish. Deep stops and shorter decompression times are characteristics of their profiles. In step with bubble and nucleation theory, these divers make their deep dive first, followed by shallower excursions. A typical series might start with a dive to 220 fsw followed by two dives to 120 fsw and culminate in three or four more excursions to less than 60 fsw. Often little or no surface intervals are clocked between dives. Such types of profiles literally clobber conventional GM tables but with proper reckoning of bubble and phase mechanics acquire some credibility. With ascending profiles and suitable application of pressure, gas seed excitation and bubble growth are likely constrained within the body's capacity to eliminate free and dissolved gas phases. In a broad sense, the final shallow dives have been tagged as prolonged safety stops and the effectiveness of these procedures has been substantiated in vivo (dogs) by Kunkle and Beckman [25]. In-water recompression procedures similar to the Australian regimens complement Hawaiian diving practices for all the same reasons. Australian and Hawaiian diving practices ushered in a new era of diving practices, especially deep stops and related protocols. And this diving was real world and certainly not academic in scheduling,

The early thermodynamic model (TM) of Hills played heavily in analyses of the these dives as published and reported in excellent sources [2, 4, 22–25]. Profile and model comparisons can be seen in Hills book [2]. While not a true bubble

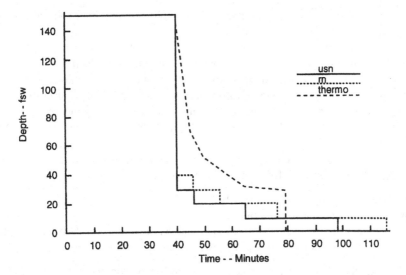

Fig. 1 Thermodynamic staging versus USN and RN staging

model per se, the TM set the stage for deep stop BM models to follow. The TM is complex in application and has not been encoded into any commercially marketed dive computers or diveware except the CCPlanner. For completeness and so it is not lost in time, the TM was described earlier. Figure 1 contrasts TM staging against RN and USN protocols of the time (1960s) and the deep stop features of the TM are evident [8–13].

Open Ocean Deep Stop Trials

Starck and Krasberg [26] in open ocean conducted a series of important deep stop tests. In deep waters in over 800 dives for up to an hour and down to 600 fsw they recorded only 4 DCS cases. Extensions to 800 fsw followed. This effort was part of a massive program to test new RB designs. The impact at the time was notable and still is today across the spectrum of diving.

Recreational 1/2 Deep Stops and Reduced Doppler Scores

Analysis of more than 16,000 actual dives by Diver's Alert Network (DAN) prompted Bennett [27] to suggest that decompression injuries are likely due to ascending too quickly. He found that the introduction of deep stops, without changing the ascent rate, reduced high bubble grades to near zero from 30.5/stops.

He concluded that a deep stop at half the dive depth should reduce the critical fast gas tensions and lower the DCS incidence rate.

Earlier Marroni [28] concluded studies with the DSL European sample with much the same thought. Although he found that ascent speed itself did not reduce bubble formation, he suggested that a slowing down in the deeper phases of the dive (deep stops) should reduce bubble formation. He has been conducting further tests along those lines. The Bennett and Marroni findings were formally incorporated into NAUI Recreational Air and Nitrox Tables [29] in 2008 for both conventional USN and No Group RGBM Tables.

The recreational regimen adopted for nonstop and light decompression diving in the NAUI Tables is straightforward and simple:

- make a 1 min stop at 1/2 bottom depth
- make a 2 min stop at 1/4 bottom depth
- if necessary and deeper than 160 fsw make a 3 min stop at 1/8 bottom depth.
  and all 1/2 deep stops made within any requisite light decompression schedules

Shallow safety stops are also made inside the deep stop recreational regimes. Obviously shallow safety and 1/2 deep stops can overlap in the 20–30 fsw range.

As Doppler scores have mostly only been correlated with light DCS symptomology (like limb bends) the above regimen appears a sane strategy for recreational air and nitrox diving. Either way though, most divers would prefer to keep Doppler scores minimal for any kind of diving and bubble reducing protocols are always prudent [16, 17, 19, 21].

Trondheim Pig Decompression Study

Brubakk and Wienke [30] found that longer and shallower decompression times are not always better when it comes to bubble formation in pigs. They found more bubbling in chamber tests when pigs were exposed to longer but shallower decompression profiles, specifically staged shallow decompression stops produced more bubbles than slower (deep) linear ascents. RGBM model predictions of separated phase under both types of decompression staging correlated with medical imaging.

Duke Chamber Experiments

Bennett and Vann [31] used a linear diffusion (TM) model to improve stops in a dive to 500 fsw for 30 min which proved DCS free in chamber tests at Duke. The early TM of Hills however at the time suggested dropout in the shallow zone which was troublesome in tests and was later modified with additional shallow decompression time. BMs today while making necessary model deep stops also require time in the shallow zone (10–$30\ fsw$). Unfortunately, premature dropout in the shallow zone may have discredited deep stop models especially the TM for the wrong reasons.

ZHL and RGBM DCS Computer Statistics

An interesting study by Balestra [32] of DAN Europe (DSL) centered on DCS incidence rates on dissolved gas, shallow stop (ZHL) computers versus bubble model, deep stop (RGBM) computers. In 11,738 recreational dives, a total of 181 DCS cases were recorded and were almost equally divided between the ZHL and RGBM computers, that is, the ZHL incidence rate was 0.0135 and the RGBM incidence rate was 0.0175. Clearly both RGBM and ZHL computers are nominally safe at roughly the 1% DCS level in this wet test. DCS rates for both computers, however, are higher than published DAN recreational rates nearer 0.1% or so.

Profile Data Banks

Computer downloaded profiles serve as a global set of diving outcomes across all diving venues and provide statistical data that can never be reproduced in chambers, wet pods and open ocean testing because of cost and diversity. The low incidence rates in these collections suggest that divers on computers are not at high risk, DCS and oxtox spikes are non existent, models and algorithms are safe and divers are using them sensibly.

- **LANL DB** With a low prevalence of deep stop DCS hits in the LANL DB (28/3569), some regard the downloaded profiles as a wet test of actual OC and RB diving. While low incidence rates are beneficial to divers, low incidence rates make statistical analysis more difficult. With the incidence rate so low in the LANL DB, the *low p* Weibull function is a more economical descriptor of the bends distribution than the binomial distribution. The DCS incidence rate in the LANL DB is 28/3569.
- **DAN DB** Like the LANL DB the massive DAN DB can also be regarded as an extended wet test for air and nitrox diving. Mixed gas and altitude profiles are also being included at last reading. With a low incidence rate (80/18745) the DAN DB underscores the relative safety of recreational air and nitrox diving. Both GM and BM profiles are stored.

In addition to DCS outcomes, broadband analysis of PDE and DSL data shows some interesting features:

- models do not always extrapolate outside their calibration (data) points
- probabilistic techniques coupled to real models are useful vehicles for diver risk estimation
- dive conditions (environmental stresses) may significantly affect risk
- body mass index (BMI) often correlates with DCS risk, particularly for older and overweight divers
- human characteristics such as age, sex and certification level affect the likelihood of diving morbidity and fatality
- leading causes of morbidity and mortality in diving are drowning, near drowning, barotrauma during ascension and then DCS
- only 2% of recreational divers use tables for dive planning with the rest relying on dive computers
- nitrox diving is exploding in the recreational sector

VVAL18 Evaluation

The recent VVAL18 compilation [33] by USN investigators is both a massive undertaking and update to USN diving data and operational protocols. With a data base of many 1000s of dives, Thalmann correlated a linear-exponential model (LEM) to data and all present USN Tables and protocols are based on it. Some impetus for this undertaking was a need for safe constant pp_{O_2} staging regimens after traditional GM approaches proved unsafe. The USN LEM is an exponential gas uptake and linear gas elimination model where as traditional GM algorithms are exponential in both gas uptake and elimination. Linear gas elimination is slower than exponential gas elimination. In marketed dive computers today, the same effect of slowing outgassing can be accomplished by increasing tissue halftimes whenever the instantaneous tension, Π, is greater than ambient pressure, p_a, in what is called the asymmetric tissue model (ATM) [4]. Greater tissue halftimes slow outgassing resulting in increased dissolved gas loadings and subsequent decompression debt. Asymmetric gas uptake and elimination can be applied to any GM or BM protocol with the same result. In the case of BM algorithms, slower outgassing contributes to bubble growth with increasing decompression requirements. A later impetus was the need for a USN dive computer for SEAL Team operations and recorded higher incidence of DCS in very warm waters.

NEDU Deep Stop Tests

Of recent interest are the deep stop (wet pod) air trials conducted at NEDU by the USN [34]. The trials generated discussion and some controversy in the technical diving community. Some 100+ air dives to 170 fsw for 30 min were staged in a wet pod at NEDU with a DCS incidence rate of $r = 5.5\%$. The staging model used was called the bubble volume model (BVM3) keyed to dissolved gas volume content not bubbles. Investigators concluded that deep stops were riskier than shallow stops though the profile tested greatly differed from BM schedules technical divers might use. This is seen in Fig. 2 which contrasts some standard BM schedules against the USN schedule. Deep air diving is also controversial as commercial operations mandate helium for depths beyond 150 fsw. But of course, the USN has its own reasons though not clear to all at the time. Some World Navies still dive deep air while commercial diving operations employ trimix beyond 150 fsw or so. Air risk increases 5–8 times beyond 160 fsw according to some commercial reports [4, 35]. For comparison purposes, Table 2 of chapter "Computer Diveware" tabulates GM and BM profiles with risk estimates against the actual wet pod profile and DCS hit rate of $r = 5.5\%$.

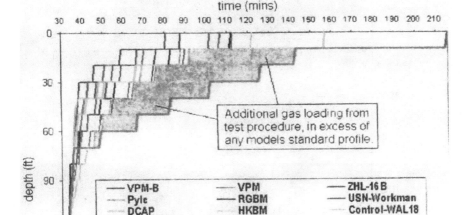

Fig. 2 NEDU deep stop test profile and model comparisons

French Navy Deep Stop Tests

Similar to the NEDU tests, the French Navy tested somewhat arbitrary deep stop schedules to $200\,fsw$ for $20\,min$ in the open ocean, that is, by inserting deep stops into their standard table schedules [35]. Of 3 profiles tested, none exhibited Grade 4 Doppler bubbles but Grade 3 bubbles were noted on deep stop schedules. The rationale is not clear for placement of deep stops but surfacing bubble counts were higher in their tests. Similar to Fig. 2 for the NEDU trials, Fig. 3 contrasts schedules for the French deep stop experiment, standard French Navy table (MN90) protocols and RGBM staging. It is strange that the deep stop profile had the longest run time

Computer Vendor and Training Agency DCS Poll

At a recent UHMS/NAVSEA Workshop deep stop statistics from dive computer Vendors and Training Agencies were presented following polling. In the anecdotal category as far as pure science and medicine they are reproduced below. The reader can take them for whatever worth but the DCS incidence rate suggested is low. That

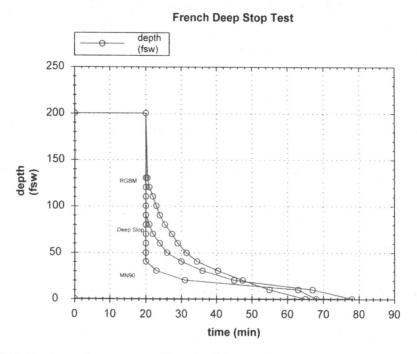

Fig. 3 French navy deep stop test profiles and model comparisons

is no surprise as DCS and oxtox spikes would likely lead to recalls and replacement units.

At that Deep Stop Workshop [35] in Salt Lake City in 2009, Training Agencies, decompression computer Manufacturers and dive software Vendors were queried prior to the Workshop for estimated DCS incidence rates against total dives performed with deep stops. Both recreational and technical diving are lumped together in their estimates (guesstimates if you like). Keep in mind that polling does not involve controlled testing and only echoes what the Agencies, Manufactures and Vendors glean from their records and accident reports. Both GM and BM algorithms with deep stops are tallied. A rough compendium of the poll is tabulated below as DCS incidences/total dives:

- Deep Stop Decompression Meters – Suunto, Mares, Dacor, Hydrospace, UTC, Atomic Aquatics, Cressisub report 47/4,000,000 with 750,000 meters marketed
- Deep Stop Software Packages– Abyss, GAP, NAUI GAP, ANDI GAP, Free Phase RGBM Simulator, NAUI RGBM Dive Planner, RGBM Simulator, CCPlanner report 68/920,000 with 30,000 CDs marketed
- Deep Stop Agency Training Dives – NAUI, ANDI, FDF, IDF report 38/1,020,000 as open water training activities
- Commercial Diving – under review with preliminary analysis

Broadly, the tally is 153/6,000,000, probably on the conservative side and slightly limited in participation.

Training Agency Testing and Standards

Some Agencies have conducted wet tests and implemented deep stop protocols into training regimens formally or optionally (NAUI, PADI, GUE, TDI, ANDI, IANTD). This is described in the Deep Stop Workshop Proceedings [35] in completeness and we only summarize a few other points in addition to the above poll. Prior to the introduction of deep stops Training Agencies relied on GM approaches in training divers and instructors with successful and safe results. The ZHL and USN table and computer implementations were mainstays in their training. When deep stop protocols entered the training scene in the 1990s, some Agencies (rather quickly) adopted a look and see attitude while applying their own testing and modified training regimens to BM algorithms, mostly VPM and RGBM. Without DCS and oxtox issues with deep stops, deep stop training standards were then drafted and implemented. As far as training regimens go, the following summarizes training standards for some well know US Agencies:

- **NAUI:** a recreational and technical Training Agency using RGBM tables, meters and linked software
- **PADI:** a recreational and technical Training Agency using DSAT tables, meters and software with deep stop options

- **SSI:** a recreational Training Agency using modified USN tables
- **ANDI:** a technical Training Agency using RGBM table, meters and diveware
- **SDI/TDI/ERDI:** a recreational and technical Training Agency using USN tables, computers and commercial diveware
- **IANTD:** a recreational and technical Training Agency employing the ZHL and VPM tables, computers and software
- **GUE:** a technical Training Agency that uses ZHL and VPM tables, computers and software

Training Agencies using USN and ZHL protocols for technical instructor often couple GFs to dive planning. Some using tables have modified times and repetitive groups to be more conservative. CMAS affiliated Training Agencies are free to choose their tables, meters and software for training. FDF and IDF employ RGBM tables, meters and software. An important thing here to mention is that regardless of training standards, tables, meters and software the training record of all collectively is safe and sane.

The decompression problem is two-pronged as depicted in Fig. 4. GM algorithms try to deal with dissolved gases by bringing the diver to the shallow zone while BM algorithms try to minimize bubble growth by staging divers in deep zones. Both seem to work and have their merits from diver vantage points and as implemented and used in dive computers.

Risk estimators follow, particularly *end-of-dive* (EOD) and *on-the-fly* (OTF) risk estimators, and are applied to profiles resident in the LANL DB as well as to nonstop diving. Risk estimators have not yet been incorporated into dive computers and diveware. Probably it is just a matter of time and is needed.

Fig. 4 Dual phase staging options

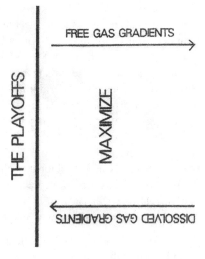

References

1. Brubakk AO and Neuman TS, *Physiology and Medicine of Diving*, Saunders Publishing, 2003, London.
2. Hills BA, *Decompression Sickness: The Biophysical Basis of Prevention and Treatment*, Wiley and Sons Publishing, 1977, Bath.
3. Bove AA and Davis JC, *Diving Medicine*, Saunders Publishing, 2004, Philadelphia.
4. Wienke BR, *Science of Diving*, CRC Press, 2015, Boca Raton.
5. Joiner JJ, *NOAA Diving Manual: Diving for Science and Technology*, Best Publishing, 2001, Flagstaff.
6. Wienke BR, *Biophysics and Diving Decompression Phenomenology*, Bentham Science Publishers, 2016, Sharjah.
7. Schreiner HR and Hamilton RW, *Validation of Decompression Tables*, Undersea and Hyperbaric Medical Society Workshop, 1989, UHMS Publication 74(VAL) 1-1-88, Washington DC.
8. Wienke BR, *Basic Decompression Theory and Application*, Best Publishing, 2003, San Pedro.
9. Wienke BR and Graver DL, *High Altitude Diving*, NAUI Technical Publication, 1991, Montclair.
10. Blogg SL, Lang MA and Mollerlokken A, *Validation of Dive Computers Workshop*, 2011, EUBS/NTNU Proceedings, Gdansk.
11. Lang MA and Hamilton RW, *AAUS Dive Computer Workshop*, University of Southern California Sea Grant Publication, 1989, USCSG-TR-01-89; Los Angeles.
12. Vann RD, Dovenbarger J and Wachholz, *Decompression Sickness in Dive Computer and Table Use*, DAN Newsletter 1989; 3–6.
13. Westerfield RD, Bennett PB, Mebane Y and Orr D, *Dive Computer Safety*. Alert Diver 1994; Mar-Apr: 1–47.
14. Workman RD, *Calculation of Decompression Schedules for Nitrogen-Oxygen and Helium-Oxygen Dives*, 1965, USN Experimenal Diving Unit Report, NEDU, 6–65, Washington DC.
15. Buhlmann AA, *Decompression: Decompression Sickness*, Springer-Verlag Publishing, 1984, Berlin.
16. Yount DE and Hoffman DC, *On the Use of a Bubble Formation Model to Calculate Diving Tables*, Aviat. Space Environ. Med. 1986; 36: 149–156.
17. Wienke BR, *Reduced Gradient Bubble Model in Depth*, CRC Press, 2003, Boca Raton.
18. Wienke BR. *Computer Validation and Statistical Correlations of a Modern Decompression Diving Algorithm*. Comp. Biol. Med. 2010; 40: 252–270.
19. Wienke BR and O'Leary TR, *Diving Decompression Models and Bubble Metrics: Dive Computer Syntheses*. Comp. Biol. Med. 2009; 39: 309–311.
20. Wienke BR, *Deep Stop Model Correlations*, J. Bioeng. Biomed. Sci. 2015; 5: 12–18.
21. Kahaner D, Moler C and Nash S, *Numerical Methods and Software: CLAMS*, Prentice-Hall, 1989, Engelwood Cliffs.
22. Le Messurier DH and Hills BA, *Decompression Sickness: A Study of Diving Techniques in the Torres Strait*, Hvaldrat. Skrif. 1965; 48, 54–84.
23. Strauss RH, *Bubble Formation In Gelatin: Implications for Prevention Of Decompression Sickness*, Undersea Biomed. Res. 1974; 1, 169–174.
24. Farm FP, Hayashi EM and Beckman EL, *Diving and Decompression Sickness Treatment Practices Among Hawaii's Diving Fisherman*, University of Hawaii Sea Grant Report, 1986, UNIHI-SEAGRANT-TP-86–01, Honolulu.
25. Kunkle TD and Beckman EL, *Bubble Dissolution Physics and the Treatment of Decompression Sickness*, Med. Phys. 1983; 10, 184–190.
26. Krasberg A, *Saturation Diving Techniques*, Proceedings Fourth International Congress Biometrology, Rutgers University Press, 1966, New Brunswick.
27. Bennett PB, Marroni et al, *Effect of Varying Deep Stop Times and Shallow Stop Times on Precordial Bubble Scores After Dives to 35 msw*, Undersea Hyper. Med. 2007; 31, 399–406.

28. Marroni A, Bennett PB et al *A Deep Stop During Decompression from 82 fsw Significantly Reduces Bubbles and Fast Tissue Tensions*, Undersea Hyper. Med. 2004; 31, 233–243.
29. O'Leary TR, *NAUI Technical Diving Manual*, 2011, NAUI Worldwide Publication, Tampa.
30. Brubakk AO, Amtzen AJ, Wienke BR and Koteng S, *Decompression Profile and Bubble Formation after Dives with Surface Decompression: Experimental Support for a Dual Phase Model of Decompression*, Undersea Hyper. Med. 2003; 30, 181–193.
31. Bennett PB and Vann RD, *Workshop on Decompression Procedures for Depths in Excess of 400 fsw*, 1975, UHMS Publication 98 (VAL), Washington DC.
32. Balestra C, *Validation of Dive Computers Workshop*, 2010, DAN-DSL Proceedings, Gdansk.
33. Thalmann ED, *Phase II Testing of Decompression Algorithms for Use in the US Navy Underwater Decompression Meter*, 1984, NEDU Report 1–84, Panama City.
34. Doolette DJ, Gerth WA and Gault KA, *Redistribution of Decompression Stop Time from Shallow to Deep Stops Increases Incidence of Decompression Sickness in Air Decompression Dives*, 2011, NEDU Report 2011–06, Panama City.
35. Bennett PB, Wienke BR and Mitchell S, *Decompression and the Deep Stop Workshop*, 2008, UHMS/NAVSEA Proceedings, Salt Lake City.

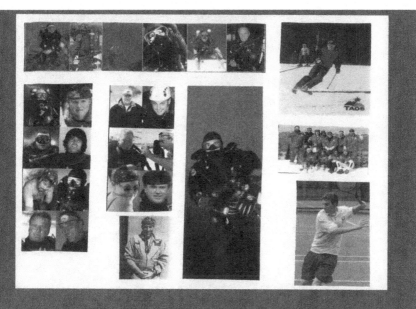

Risk Estimators

Risk estimation [10–13] on the fly (OTF) or end of dive (EOD) is not yet implemented in dive computers nor planning software. The following suggests appropriate methodology for implementation of both. As dive computers working in the recreational (air and nitrox) depth regime, $d < 130\,fsw$ roughly, use GM models for speed and simplicity and dive computers working in the technical (mixed gases and decompression) depth regime, $d > 130\,fsw$, employ BM models, we will use use GM risk functions in comparative applications for shallow recreational diving, $d < 130\,fsw$, and BM risk functions in comparative applications for deep and decompression technical diving, $d > 130\,fsw$. The USN and ZHL models [14, 15] are GMs and the VPM and RGBM models [3, 4] are BMs and will be employed for staging calculations. These results have also been published. Applying these techniques to the profiles generated in the previous wet and dry tests would be an interesting exercise using existing diveware as detailed in the next Section.

End of Dive Risk Estimator (EOD)

In performing risk analysis with the LANL DB, the tissue gradient is useful. As seen before, [1–9], the gradient is cast into normalized risk function, ρ, form,

$$\rho(\kappa, \omega, t) = \kappa \left[\frac{\Pi(t) - P(t)}{P(t)} \right] - \kappa \, exp\,(-\omega t)$$

with $\Pi(t)$ and $P(t)$ total tissue tension and ambient pressure in time, t, respectively. Risk is quantified by the difference between total tissue tension and ambient pressure divided by ambient pressure summed over time. Risk increases with increasing tissue tension and decreasing ambient pressure and increasing time. The approach was used before for overall dive risk estimation [1, 6, 7]. An asymptotic exposure limit is used in the risk integrals, that is, $t_{mx} = 48\,hrs$ after surfacing across all compartments, τ, in time, t,

$$1 - r(\kappa, \omega) = exp\left[-\int_0^{t_{mx}} \rho(\kappa, \omega, t)dt\right]$$

with $r(\kappa, \omega)$ the usual cumulative risk after time, t. The first term in the risk function, ρ, links to dynamical supersaturation in the models while the second term is a smoothing function over dive time. To estimate κ and ω within maximum likelihood (ML), a Weibull-Levenberg-Marquardt (WLM) [4, 18] *low p* package was employed ($SNLSE$, Common Los Alamos Mathematical And Statistical Library) [21], a non linear least squares data fit (NLLS) to an arbitrary exponential function (minimization of variance over 3569 data points with $L2$ error norm). The computational program is straightforward but massive. Across all tissue compartments, τ, the maximum value of the gradient is cumulated in the risk integral every 5–$10\,sec$ until surfacing and across all profiles. A resulting 3659×3659 matrix is stored for further manipulation, inversion and minimization. Across GM and BM algorithms (US Navy, ZHL, VPM, RGBM) and using Table 2 of chapter "Computer Profile Data", there then obtains a range for the fit parameters [19],

$$\kappa = 0.698 \pm 0.283\,min^{-1}$$

$$\omega = 0.810 \pm 0.240\,min^{-1}$$

Mathematical and computing details are given elsewhere for EOD risk estimation [6, 22] and not repeated. Recall that EOD risk estimates extend out to $2\,hrs$ (t_{mx}) after the dive. Some earlier EOD risk estimates follow [20] for select profiles and staging for completeness. Risk estimators are denoted, r_{GM} and r_{BM} to distinguish algorithms. For numerics the ZHL, USN, VPM and RGBM models were used in the following but the results are trendwise generic for GM and BM classes of algorithms. In the following depth, $d = 130\,fsw$, separates GM and BM applications. These EOD risk estimators will serve as surfacing bootstraps for OTF risk estimators in the next section. There will be 3659 EOD risk estimates to link to the surfacing 3659 OTF risk estimators for profiles in the LANL DB using NLLS techniques.

Test Profiles and EOD Risk

Following examples are taken from the LANL DB and have been discussed and published earlier [4, 22] with regards to models, staging comparisons, deep and shallow stops, tests and data.

1. **Deep OC Trimix Dive** Consider a deep trimix dive with multiple gas switches on the way up. This is a risky technical dive performed by seasoned professionals. Table 1 contrasts stop times for two gas choices at the $100\,fsw$ switch. The dive is a short $10\,min$ at $400\,fsw$ on TMX 10/65 with switches at 235, 100 and $30\,fsw$. Descent and ascent rates are 75 and $25\,fsw/min$. Obviously, there are many other choices for switch depths, mixtures and strategies. In this comparison, oxygen fractions were constant in all mixes at all switches. Differences between a nitrogen or a helium based decompression strategy, even for this short exposure, are nominal. Such usually is the case when oxygen fraction is held constant in helium or nitrogen mixes at the switch.

   Comparative and anecdotal diver reports suggest that riding helium to the $70\,fsw$ level with a switch to EAN50 is a good strategy, one that couples the benefits of well being on helium with minimal decompression time and stress following isobaric switches to nitrogen. Shallower switches to enriched air also work with only a nominal increase in overall decompression time, but with deeper switches off helium to nitrox a source of isobaric counterdiffusion (ICD) issues that might best be avoided. Note the risk, r_{RGBM}, for the helium strategy, TMX 40/20 at $100\,fsw$, is slightly safer than the nitrogen strategy, EAN40 at $100\,fsw$, but in either case the risk is high.

   The logistics of such deep dives on OC are formidable for both diver and support crew (if any). The number of stage bottles (decompression tanks) is forbidding for a single diver, of course, but surface support teams, themselves at high risk for placing bottles on a line at depth, can effect such a dive. These support teams are vested with immense responsibility for diver safety.

2. **Hydrospace EXPLORER Extreme RB Profile** Table 2 is a deep RB dive downloaded off the Hydrospace EXPLORER computer. From a number of corners reports of $400\,fsw$ dives on rebreather systems are becoming commonplace. Consider this one to $444\,fsw$ for $15\,min$. Diluent is TMX 10/85 and pp_{O_2} setpoint is $1.1\,atm$. From a decompression standpoint, rebreather systems are the quickest and most efficient systems for underwater activities. The higher the pp_{O_2}, the shorter the overall decompression time. That advantage, however, needs to be played off against increasing risks of oxygen toxicity as oxygen partial pressures increase, especially above $1.4\,atm$. The higher percentage of oxygen and lower percentage of inert gases in higher pp_{O_2} setpoints of closed circuit rebreathers (CCRs) results in reduced risks simply because gas loadings and bubble couplings are less in magnitude and importance. This shows up in any set of RB comparative pp_{O_2} calculations as well as in OC versus RB risk estimates. Risk associated with this $444\,fsw$ dive is less than a similar dive on trimix to roughly same depth for shorter time, that is, looking at Table 1. Certainly, this reduction relates to the higher oxygen fraction in RB systems.

While the above approach across all profiles using DCS outcomes is amenable to implementation in dive planning software with suitable computer processor speeds and storage resources (PCs, Workstations, Clusters, Mainframes, etc.), it is not always optimal in present generation underwater dive computers. They are

Table 1 Comparative
helium and nitrogen gas
switches and risk

Depth (*fsw*)	Time (*min*)	Time (*min*)
	TMX 10/65	TMX 10/65
400	10.0	10.0
260	1.5	1.5
250	1.0	1.0
240	1.0	1.0
	TMX 18/50	TMX 18/50
230	0.5	0.5
220	0.5	0.5
210	0.5	0.5
200	0.5	0.5
190	1.0	1.0
180	1.5	1.5
170	1.5	1.0
160	1.5	1.5
150	1.5	2.0
140	2.0	1.5
130	2.0	2.5
120	4.0	4.0
110	4.5	4.0
	TMX 40/20	EAN40
100	2.5	2.0
90	2.5	2.0
80	2.5	2.0
70	5.0	4.0
60	6.5	5.5
50	8.0	6.5
40	9.5	7.5
	EAN80	EAN80
30	10.5	10.5
20	14.0	14.0
10	21.0	20.5
Run time	123.0	116.0
	$r_{RGBM} = 6.42\%$	$r_{RGBM} = 6.97\%$

more limited in computing speed and storage capabilities. Divers also want to
know risks during a dive not just at the end. So a modified approach using the
data in Tables 1, 2, and 3 of chapter "Computer Profile Data" is suggested using
permissble supersaturation during the dive. Consider the following model risk
estimators easily generated on the fly by present dive computers. Unlike the previous
cumulative estimators these can be viewed underwater as the dive progresses.

Table 2 Extreme RB dive
and risk

Depth (fsw)	Time (min)	Depth (fsw)	Time (min)
444	15.	150	2.0
290	0.5	140	2.0
280	0.5	130	2.0
270	0.5	120	2.5
260	0.5	110	3.0
250	0.5	100	3.5
240	0.5	90	4.0
230	1.0	80	4.5
220	1.0	70	5.0
210	1.0	60	7.0
200	1.0	50	7.5
190	1.5	40	8.0
180	1.5	30	12.5
170	1.5	20	14.0
160	1.5	10	15.5
			$r_{VPM} = 6.79\%$

On the Fly Risk Estimator (OTF)

As DCS outcomes for excursions from any point on a dive to the surface or elsewhere above the diver are unknown the approach used for EOD risk is not portable directly to OTF risks. The foregoing does suggest another computational approach at any depth in terms of model limit points above the diver, specifically, critical gradients, G and H, for GM and BM models respectively.

For GM risk, we have,

$$r(\alpha, \beta, \epsilon, t) = \alpha\, exp \left[-\left(\frac{\Pi(t) - P(t) - G(t)}{P(t)}\right)\right] + \beta[1 - exp\,(-\epsilon t)]$$

with published permissible gradient, G, in the M-value picture (USN),

$$G = M - P$$

and similarly published gradient, G, in the Z-value picture (ZHL),

$$G = Z - P$$

Accordingly, for BM algorithms, it follows,

$$r(\alpha, \beta, \epsilon, t) = \alpha\, exp \left[-\left(\frac{\Pi(t) - P(t) - H(t)}{P(t)}\right)\right] + \beta[1 - exp\,(-\epsilon t)]$$

One published permissible BM bubble-tissue gradient, H, is averaged over the bubble seed distribution (RGBM),

$$H = 2\gamma\zeta \int_{r_c}^{\infty} \frac{exp\ [-\zeta(r - r_c)]}{r} dr$$

with surface tension, γ, given by,

$$2\gamma = 44.7 \left[\frac{P}{T}\right]^{1/4} + 24.3 \left[\frac{P}{T}\right]^{1/2}\ dyne/cm$$

for T temperature (^oK), P ambient pressure (fsw) and r_c critical radius (μm) for ζ a fitted constant (order 0.7 μm^{-1}) for the bubble distribution with nitrogen,

$$r_c = 0.007655 + 0.001654 \left[\frac{T}{P}\right]^{1/3} + 0.041602 \left[\frac{T}{P}\right]^{2/3}$$

and for helium,

$$r_c = 0.003114 + 0.015731 \left[\frac{T}{P}\right]^{1/3} + 0.025893 \left[\frac{T}{P}\right]^{2/3}$$

Another published BM permissible bubble-tissue gradient, H, takes the gel form (VPM),

$$H = \frac{2\gamma\ (\gamma_c - \gamma)}{\gamma_c r_c} = \frac{11.01}{r_c}\ (fsw)$$

for γ and γ_c film and surfactant surface tensions, that is, $\gamma = 17.9\ dyne/cm$ and $\gamma_c = 257\ dyne/cm$ with critical bubble radius r_c in μm given by,

$$\frac{1}{r_c} - \frac{1}{r_i} = \frac{P - P_i}{2\gamma}$$

with $r_i = 0.6\ \mu m$ at sea level, that is, $P_i = 33\ fsw$. The BM permissible gradients range at 10–40 fsw roughly.

The OTF functions are quantified by the difference between existing and permissible supersaturation divided by ambient pressure. Risk increases with increasing difference between existing and permissible supersaturation and decreasing ambient pressure. First terms are measures of permissible supersaturation differences and second terms are overall smoothing functions that increase with dive time, t. Similarly, but not integrated over run time, we define the instantaneous risk function, r, for ascents above the diver to arbitrary depths with critical parameters, G and H, and its complement, ρ,

$$\rho(\alpha, \beta, \epsilon, t) = 1 - r(\alpha, \beta, \epsilon, t)]$$

as OTF risk estimators depending on instantaneous depth, d, final ascent level, d_0, bottom time, t_b, and dive run time, t. In analogy with the EOD compilation, the maximum value of the risk function across all tissue compartments, τ, is tallied and used. This occurs with the (ascent) controlling tissue compartment at the end of nonstop bottom time or level decompression stop time with corresponding tissue tension, Π.

As data for OTF risk estimation does not exist, we will rely on an extrapolation scheme that fits the OTF risk estimator close to the surface to the EOD risk estimator after surfacing for all the profiles in the LANL DB using standard NLLS software. This is a task requiring LANL supercomputers with teraflop speeds (10^{12} floating point operations per second) and fast access mass storage accommodating a 3569×3569 matrix for NLLS inversion. What this amounts to is fitting the OTF risk function at the end of the last decompression stop or NDL for nonstop diving to the EOD risk estimator after surfacing at time, t_{mx}, across all 3569 profiles,

$$r(\alpha, \beta, \epsilon, t_{mx}) = r(\kappa, \omega, t_{mx})$$

with EOD risk estimator computed for each profile using,

$$\kappa = 0.698$$

$$\omega = 0.810$$

and α, β and ϵ then extracted in the NLLS fit to $r(\kappa, \omega, t_{mx})$. The resulting OTF risk functions are then used to estimate OTF risks at any point, d_0, above the diver with the surfacing case, $d_0 = 0$, the focus here. Obviously, for points above the diver but below the surface, risk decreases compared to surfacing risk. For GM algorithms, we obtain using the ZHL,

$$\alpha = 0.350 + 0.00125 \, (d - d_0) \pm 0.081$$

$$\beta = 0.025 \pm 0.004$$

$$\epsilon = 1/t_b \pm 0.106 \, min^{-1}$$

For BM algorithms we employ the RGBM with results,

$$\alpha = 0.550 + 0.00118 \, (d - d_0) \pm 0.053$$

$$\beta = 0.022 \pm 0.005$$

$$\epsilon = 1/t_b \pm 0.079 \, min^{-1}$$

The critical parameters, G and H (permissible tissue and bubble supersaturation gradients) are evaluated at the ascent point (d_0). Possible tissue outgassing and bubble growth during the ascent are included in the analyses assuming an ascent rate of 30 fsw/min. In GM staging, tissues likely outgas during ascent, reducing tissue tensions and risk. In BM staging, bubbles grow on ascent when not controlled by stops and risk increases. For surfacing ascents from any point on the dive, $d_0 = 0$. The risk for GM algorithms increases as the difference between actual tissue tension and critical tension at any point on the dive increases. For BM algorithms, risk increases as the difference between actual supersaturation and permissible bubble supersaturation increases.

Ingassing and outgassing during ascents and descents are incorporated easily into the tissue equations by assuming ambient pressure, p_a, is changing in time. For assumed linear ascent rate, v, we have,

$$p_a = p_0 - vt$$

with speed, v, positive for descents and negative for ascents (convention). The corresponding tissue equation becomes,

$$\frac{\partial p}{\partial t} = -\lambda(p - p_0 + vt)$$

with straightforward solution, $p = p_i$, at, $t = 0$,

$$p = p_0 + (p_i - p_0 + v/\lambda) \, exp \, (-\lambda t) - vt - v/\lambda$$

At initial time, $t = 0$, or stationary diver, $v = 0$, the equation reduces to the usual form. For long ascents or descents, tissue loadings become important and changes in gas tensions, p, need be included in calculations of risk. If omitted on descent tissue tensions are smaller and if omitted on ascent tissue tensions are larger than estimated with the static equation. Effects are seen in both GM and BM algorithms. For GM algorithms changes in gas loadings with ascent are fairly simple as seen above. For BM algorithms the situation is more complex in that changes in gas loadings on ascent affect gas diffusion across bubble interfaces with bubble behavior additionally becoming a matter of surface tension and bubble size. In the following, gas loadings and bubble changes are tracked during ascents and descents.

On the fly risk estimates for various trimix, nitrox and heliox dives follow. In all cases, OTF surfacing risks at the end of the decompression stop time or NDL are tabulated using nominal values for α, β and ϵ listed.

Test Profiles and OTF Risk

As with EOD profiles, the following are examples taken from the LANL DB having been discussed and published earlier [4, 18, 20] with regards to models, staging comparisons, deep and shallow stops, tests and data.

1. **Recreational Nonstop Air Diving** Many hundreds of air dives were analyzed by the USN permitting construction of decompression schedules with 95% and 99% confidence (5% and 1% bends probability). These tables were published by USN investigators [1, 7] and Table 2 tabulates the corresponding nonstop time limits ($\sigma = 0.05, 0.01$) and also includes the old USN (Workman) limits for comparison in the fourth column. They date back to the 1950s. Later re-evaluations of the standard set of nonstop time limits estimate a probability rate of 1.25% for the limits. In actual usage, the incidence rates are below 0.01% because users do not dive to the limits generally. In the last columns are listed risk estimates, r_{ZHL} and r_{RGBM}, for the 1% DCS probability USN limits, $\sigma = 0.01$, using on the fly estimators. Again, $d_0 = 0$ in the nonstop case for the conservative NDLs. The risk estimates in the last two columns include outgassing during ascent with ascent rate of 30 fsw/min and ingassing during descent with descent rate of 60 fsw/min.

   It is clear in Table 3 that the USN 1% and corresponding on the fly risks, r_{ZHL} and r_{RGBM}, are very close. Both ZHL and RGBM risks in Table 3 are slightly larger and so more conservative in dive computer and diveware applications. As noted before, GM and BM algorithms overlap in the nonstop diving limit because phase separation is minimal in BM algorithms [6, 19]. Over nonstop air diving to recreational limits, we have across the ZHL,

$$1.60\% \leq r_{ZHL} \leq 2.08\%$$

with for the RGBM,

$$r_{RGBM} \leq r_{ZHL}$$

Table 3 Nonstop air limits and risk

Depth d (fsw)	Nonstop limit t_n (min) $\sigma = 0.05$	Nonstop limit t_n (min) $\sigma = 0.01$	Nonstop limit t_n (min) USN	Risk r_{ZHL}	Risk r_{RGBM}
30	240	170		0.0160	0.0160
40	170	100	200	0.0162	0.0161
50	120	70	100	0.0166	0.0163
60	80	40	60	0.0166	0.0165
70	80	25	50	0.0177	0.0169
80	60	15	40	0.0174	0.0173
90	50	10	30	0.0179	0.0178
100	50	8	25	0.0190	0.0184
110	40	6	20	0.0199	0.0192
120	40	5	15	0.0204	0.0196
130	30	4	10	0.0208	0.0200

for USN corresponding 1% risk. The decrease in nonstop time limits as risk drops into the 1% range is interesting compared to early USN compilations (Workman). This was run with GAP and CCPlanner.

2. **Deep Trimix OC Dive** The following is a deep TMX 16/46 dive with helium-oxygen mirroring and constant nitrogen gas fraction on all ascent switches, that is, $f_{N_2} = 0.38$ until a final switch to EAH80 at 20 fsw. The switches are TMX 18/44 at 220 fsw, TMX 20/42 at 140 fsw, TMX 22/40 at 80 fsw and EAH80 at 20 fsw. This is an optimal strategy on many counts. Table 4 lists pertinent dive variables and corresponding on the fly risks for immediate (emergency) surfacing ascent at any of the stops. The variable $psat$ is the permissible RGBM supersaturation. Other entries are self explanatory.

Above and following examples are tabulated using CCPlanner with nominal parameter settings corresponding to settings in meters and software.

3. **Shallow Nitrox OC Dive** A decompression dive on EAN32 without any gas switches is analyzed in Table 5. The profile is EAN32 at 100 fsw for 65 min. Entries are the same as Table 4. The OTU and CNS entries are the full body and CNS cumulations at each level. Decompression profile and surfacing risks are listed.

Table 4   Deep trimix OC dive and risk

Depth (fsw)	Wait (min)	Tissue (min)	Tension (fsw)	psat (fsw)	pp_{O_2} (atm)	Risk r_{RGBM}
300	15.0	3.3	274.5	30.7	1.6	0.407
190	0.5	3.3	205.5	32.3	1.2	0.330
180	0.5	5.3	202.3	33.3	1.2	0.320
170	1.0	5.3	189.4	33.3	1.1	0.308
160	1.0	5.3	182.0	33.3	1.1	0.236
150	1.0	5.3	170.6	33.3	1.0	0.200
140	1.0	8.2	162.6	34.3	1.0	0.197
130	1.5	8.2	151.7	34.1	1.0	0.195
120	1.5	8.2	141.7	34.1	0.9	0.193
110	1.5	12.2	134.1	34.9	0.9	0.189
100	2.5	12.2	123.7	34.9	0.8	0.184
90	2.5	12.2	114.1	34.9	0.7	0.179
80	3.0	17.8	105.1	35.5	0.8	0.170
70	4.0	17.8	94.9	35.5	0.7	0.160
60	4.5	25.3	85.6	35.9	0.6	0.147
50	6.5	25.3	75.4	35.9	0.6	0.132
40	7.5	35.9	66.1	36.2	0.5	0.112
30	10.5	35.9	56.0	36.2	0.4	0.087
20	8.5	50.8	46.2	36.3	1.3	0.051
10	12.5	72.0	36.3	36.4	1.0	0.031
	101.5					

Table 5 Shallow nitrox OC dive and risk

Depth (fsw)	Wait (min)	Tissue (min)	Tension (fsw)	psat (fsw)	pp_{O_2} (atm)	OTU (min)	CNS	Gas (ft^3)	Risk r_{ZHL}
100	65.0	18.5	89.4	49.1	1.3	95.0	0.42	200	0.173
20	5.5	27.1	43.5	43.0	0.5	0.3	0.00	7	0.067
10	24.0	54.4	23.9	32.9	0.4	0.0	0.00	23	0.019
	99.5					95.3	0.42	230	

Table 6 Heliox RB dive and risk

Depth (fsw)	Wait (min)	Tissue (min)	Tension (fsw)	psat (fsw)	pp_{O_2} (atm)	OTU (min)	CNS	Risk r_{VPM}
344	15.0	1.9	344.1	43.6	1.0	15.0	0.04	0.206
240	0.5	3.0	269.8	43.9	1.1	0.6	0.00	0.183
230	0.5	4.7	260.5	43.4	1.1	0.6	0.00	0.184
220	0.5	4.7	252.4	43.4	1.1	0.6	0.00	0.182
210	1.0	4.7	239.2	43.4	1.1	1.2	0.00	0.178
200	1.0	4.7	230.8	43.4	1.2	1.3	0.01	0.176
190	1.0	7.0	222.8	43.1	1.2	1.3	0.01	0.177
180	1.5	7.0	210.1	43.1	1.2	2.0	0.01	0.172
170	1.5	7.0	201.0	43.1	1.2	2.0	0.01	0.170
160	1.5	7.0	191.8	43.1	1.2	2.0	0.01	0.167
150	1.5	10.2	182.6	42.8	1.2	2.0	0.01	0.165
140	2.0	10.2	171.6	42.8	1.2	2.6	0.01	0.160
130	2.0	10.2	160.7	42.6	1.2	2.6	0.01	0.156
120	1.5	10.2	152.1	42.6	1.2	2.0	0.01	0.152
110	2.5	14.5	141.3	42.4	1.2	3.3	0.01	0.147
100	2.5	14.5	131.6	42.4	1.3	3.7	0.02	0.141
90	2.5	14.5	121.9	42.4	1.3	3.7	0.02	0.134
80	3.0	20.5	111.9	42.2	1.3	4.4	0.02	0.127
70	4.0	20.5	101.6	42.2	1.3	5.9	0.03	0.116
60	4.0	20.5	91.2	42.1	1.3	5.9	0.03	0.103
50	5.0	29.1	81.4	42.0	1.3	7.4	0.03	0.090
40	5.5	29.1	71.6	41.9	1.3	8.1	0.04	0.074
30	6.5	41.2	61.6	41.7	1.3	9.6	0.04	0.053
20	8.5	41.2	51.2	41.6	1.3	12.6	0.06	0.041
10	10.0	55.2	41.3	41.4	1.3	14.8	0.07	0.033
	102.2					115.1	0.47	

4. **Heliox RB Dive** The last OTF risk example is a pure heliox CCR dive to 344 fsw for 15 min with three setpoint changes on the way up. The diluent is heliox 21/79. setpoints are 1.0 atm at the bottom, 1.1 atm at 200 fsw, 1.2 atm at 100 fsw and 1.3 atm at 30 fsw. Table 6 gives the decompression profile with risks for surfacing from stops.

Overall risks for the deep heliox CCR dive are smaller than corresponding risks for OC dives to the same depths. The higher oxygen and lower helium gas fractions in the breathing loop lower risk as requisite. Both tissue tensions and bubbles remain smaller. Said another way, RB diving is safer for given depth and time.

Equal Risk Deep and Shallow Stop Profiles

Computationally, equal risk shallow and deep stop profiles can be compared using the BM risk functions described versus a known shallow stop GM profile and its risk. The BM risk function is parameterized within the deep stop RGBM discussed while the shallow stop GM profile and risk are taken from actual tests [22], that is, an air dive to 150 fsw for 30 min. The 150/30 profile had an incidence rate of 2.1% over 96 dives (2/96). Both deep RGBM and shallow stop USN profiles are given in Table 7 for equal surfacing risk estimators, $r_{USN} = r_{RGBM} = 2.1\%$. CCPlanner with risk estimators detailed is used for analysis, encoded with both USN and RGBM staging algorithms.

Table 7 Equal surfacing risk deep and shallow stop air dives

USN depth (fsw)	Shallow stop time (min)	r_{USN}	RGBM depth (fsw)	Deep stop time (min)	r_{RGBM}
150	30	0.224	150	30	0.298
80			80	1	0.210
70			70	1	0.201
60			60	2	0.180
50			50	3	0.152
40	3	0.129	40	4	0.125
30	7	0.101	30	5	0.093
20	12	0.069	20	7	0.050
10	31	0.024	10	10	0.023
0	83	0.021	0	63	0.021

The RGBM profile was iteratively converged to 2.1% surfacing OTF risk. The iteration process was manual cranking pointing to the utility of a software package able to connect equal risk profiles in dive computers and dive planning diveware. It's coming with simple profile iteration for equality of r_{GM} and r_{BM} just a high speed computer search and converge process.

The model differences in the OTF risk estimators on the way up are obvious as are the staging protocols. At the end of the dive, OTF and EOD estimators converge as expected from the surfacing fit described. In these calculations, outgassing on the way to the surface after the $10\,fsw$ decompression stop is included for the controlling tissue compartment to give the most realistic estimates of surfacing risk. Outgassing during the ascent to the surface is what drops the risk to 2.1% at dive conclusion. As usual case with GM versus BM staging, run times are always shorter for BM versus GM staging at the same risk level, in this case, $63\,min$ versus $83\,min$. But given the relative safety record of both GM and BM computers, both model approaches work suggesting there are two ways to dive safely, *control the bubble at depth or treat the bubble in the shallows* as a number of medical observers and seasoned divers suggest. To compare apples to apples, of course, it is necessary to first normalize the risk to the same level in both models as done here [16, 17].

References

1. Brubakk AO and Neuman TS, *Physiology and Medicine of Diving*, Saunders Publishing, 2003, London.
2. Hills BA, *Decompression Sickness: The Biophysical Basis of Prevention and Treatment*, Wiley and Sons Publishing, 1977, Bath.
3. Bove AA and Davis JC, *Diving Medicine*, Saunders Publishing, 2004, Philadelphia.
4. Wienke BR, *Science of Diving*, CRC Press, 2015, Boca Raton.
5. Joiner JJ, *NOAA Diving Manual: Diving for Science and Technology*, Best Publishing, 2001, Flagstaff.
6. Wienke BR, *Biophysics and Diving Decompression Phenomenology*, Bentham Science Publishers, 2016, Sharjah.
7. Schreiner HR and Hamilton RW, *Validation of Decompression Tables*, Undersea and Hyperbaric Medical Society Workshop, 1989, UHMS Publication 74(VAL) 1-1-88, Washington DC.
8. Wienke BR, *Basic Decompression Theory and Application*, Best Publishing, 2003, San Pedro.
9. Wienke BR and Graver DL, *High Altitude Diving*, NAUI Technical Publication, 1991, Montclair.
10. Blogg SL, Lang MA and Mollerlokken A, *Validation of Dive Computers Workshop*, 2011, EUBS/NTNU Proceedings, Gdansk.
11. Lang MA and Hamilton RW, *AAUS Dive Computer Workshop*, University of Southern California Sea Grant Publication, 1989, USCSG-TR-01-89; Los Angeles.
12. Vann RD, Dovenbarger J and Wachholz, *Decompression Sickness in Dive Computer and Table Use*, DAN Newsletter 1989; 3-6.
13. Westerfield RD, Bennett PB, Mebane Y and Orr D, *Dive Computer Safety*. Alert Diver 1994; Mar-Apr: 1-47.
14. Workman RD, *Calculation of Decompression Schedules for Nitrogen-Oxygen and Helium-Oxygen Dives*, 1965, USN Experimenal Diving Unit Report, NEDU, 6-65, Washington DC.
15. Buhlmann AA, *Decompression: Decompression Sickness*, Springer-Verlag Publishing, 1984, Berlin.
16. Yount DE and Hoffman DC, *On the Use of a Bubble Formation Model to Calculate Diving Tables*, Aviat. Space Environ. Med. 1986; 36: 149-156.
17. Wienke BR, *Reduced Gradient Bubble Model in Depth*, CRC Press, 2003, Boca Raton.
18. Wienke BR. *Computer Validation and Statistical Correlations of a Modern Decompression Diving Algorithm*. Comp. Biol. Med. 2010; 40: 252-270.
19. Wienke BR and O'Leary TR, *Diving Decompression Models and Bubble Metrics: Dive Computer Syntheses*. Comp. Biol. Med. 2009; 39: 309-311.
20. Wienke BR, *Deep Stop Model Correlations*, J. Bioeng. Biomed. Sci. 2015; 5: 12-18.

21. Kahaner D, Moler C and Nash S, *Numerical Methods and Software: CLAMS*, Prentice-Hall, 1989, Engelwood Cliffs.
22. Wienke BR, *Dive Computer Profile Data And On The Fly And End Of Dive Risk Estimators*, J. Appl. Biotech. Bioeng. 2018; 5(1): 00118.

Reduced Gradient Bubble Model (RGBM)
Dive Table - Air
6,000 to 10,000 ft / 1829 to 3048 m

Reduced Gradient Bubble Model (RGBM)
Dive Table - Air
2,000 to 6,000 ft / 610 to 1829 m

Reduced Gradient Bubble Model (RGBM)
Dive Table - Air
Sea Level to 2,000 ft / 610 m

DIVE SAFETY THROUGH EDUCATION

DIVE ONE			DIVE TWO			DIVE THREE		
MAX DEPTHS		MDT	MAX DEPTHS		MDT	MAX DEPTHS		MDT
fsw	msw	minutes	fsw	msw	minutes	fsw	msw	minutes
130	40	10	80	24	30	30	9	150
120	36	13	75	23	30	30	9	150
110	33	16	70	21	40	30	9	150
100	30	20	65	20	40	30	9	150
90	27	25	60	18	55	30	9	150
80	24	30	55	17	55	30	9	150
70	21	40	50	15	80	30	9	150
60	18	55	45	14	80	30	9	150
50	15	80	40	12	110	30	9	150
40	12	110	35	11	110	30	9	150
30	9	150	30	9	150	30	9	150

This table is designed for scuba dives employing air.

Read the instructions on the back and seek proper training before using this table or compressed air. Even strict compliance with this table will not guarantee avoidance of decompression sickness.

Computer Diveware

A potpourri of software packages available on the market are described briefly. They are chosen because of their widespread use, utility, historical perspectives and diver popularity. New ones are coming online every day. They are popularly categorized as dissolved gas, dissolved gas with GFs, pseudo-bubble and bubble models. Dissolved gas, dissolved gas with GFs and pseudo-bubble models are collectively termed neo-Haldane models In neo-Haldane models, M-values and Z-values are reduced compared to the original USN and ZHL compilations [1, 2]. The RGBM and VPM [3, 4]are the only true bubble models of interest and commercially available in diveware and computers.

Packages

Online and commercially available software packages span GM and BM algorithms along with oxtox estimation and include:

1. **Free Phase RGBM Simulator** Free Phase RGBM Simulator is a software package offered by Free Phase Diving incorporating the ZHL and RGBM algorithms. Both the ZHL and RGBM algorithms are user validated and correlated with actual diving data and tests [5, 7]. The Free Phase RGBM Simulator for nominal settings is one-to-one with the published and released NAUI Technical Diving Tables [6] used to train mixed gas OC and RB divers. As such, it is a valuable training and diving tool for deep and decompression diving. No other diveware packages, excepting NAUI GAP and ANDI GAP, provide such correlation with published and user validated Dive Tables. It is also keyed to the Liquivision RGBM implementation plus a few others under construction in the Far East.

© The Author(s), under exclusive licence to Springer Nature Switzerland AG 2018   83
B. R. Wienke, T. R. O'Leary, *Understanding Modern Dive Computers and Operation*,
SpringerBriefs in Computer Science, https://doi.org/10.1007/978-3-319-94054-0_8

2. **Abyss** Abyss in the mid 1990s first introduced the full RGBM into its diveware packages. The Buhlmann ZHL model was also included in the dissolved gas package. It has seen extensive use over the past 20 *years* or so in the technical diving area. A variety of user knobs on bubble parameters and M-values permit aggressive to conservative staging in both models. Both the ZHL and RGBM have been published and formally correlated with diving data. Later, the modified RGBM with χ critical tension multipliers was incorporated into Abyss and published and correlated with data in the late 1990s, serving also as the basis for Suunto, Mares, Dacor, ConnXion, Cressisub, UTC, Mycenae, Aqwary, Hydrospace, ANO, Artisan and other RGBM dive computers. Full RGBM was first incorporated into Hydrospace computers and today in Suunto, Atomic Aquatics, Liquivision and ANO computers.

3. **VPlanner** VPlanner first introduced the VPM in the late 1990s. Based on the original work of Yount and Hoffman, the software has seen extensive use by the technical diving community. Formal LANL DB correlations of the VPM and thus VPlanner have been published [7]. User knobs allow adjustment of bubble parameters for aggressive to conservative staging. VPlanner is also used in Liquivision and Advanced Diving Corporation computers for technical diving.

4. **ProPlanner** ProPlanner is a software package that uses modified Z-values for diver staging. Buhlmann Z-values with GFs are employed with user knobs for conservancy. The model is called the VGM ProPlanner by designers. Some GFs claim to mimic the VPM. Correlations have not been formally published about VGM and ProPlanner.

5. **GAP** GAP is a software package similar to Abyss offering the full RGBM, modified RGBM with χ and Buhlmann ZHL with GFs. Introduced in the mid 1990s, it has seen extensive usage in the recreational and technical sectors. Apart from user GFs, the models and parameters in GAP have been published and correlated with diving data and profiles tested over years. Adjustable conservancy settings for all models can be selected. GAP has been keyed to Atomic Aquatics and Liquivision dive computers. Training Agency spinoffs also include ANDI GAP and NAUI GAP.

6. **DecoPlanner** DecoPlanner is a diveware package offered by the GUE Training Agency. Both the VPM and Buhlmann ZHL with GFs are available in DecoPlanner. Evolving over the past 10–15 *years*, DecoPlanner also incorporates GUE *ratio deco* ($\Pi/P \leq \xi$) approaches to modifying GFs. Nothing is published about ratio deco data correlations but both the ZHL and VPM have been correlated [7]. It has seen extensive use in the technical diving community and GUE diver training.

7. **Analyst** Analyst is a software package marketed by Cochrane Undersea Technology for PCs. It is keyed to Cochrane computers as a dive planner and profile downloader. The Cochrane family of computers use the USN LEM for recreational, technical and military applications. The LEM is a neo-Haldanian model with exponential uptake and linear elimination of inert gases. It is part of the massive USN VVAL18 project.

8. **DiveLogger** DiveLogger is linked to Ratio technical and recreational computers. Ratio computers provide GPS and wireless connectivity and offer the ZHL and VPM algorithms to divers. Dive planning and profile downloading capabilities are included in the diveware package. As mentioned, both VPM and ZHL have been correlated with data.

9. **DiveSim** DiveSim is a UDI software package for dive planning and profile downloading. UDI computers and diveware employ the correlated RGBM for air and nitrox. The software packages also includes diver to diver, diver to surface, GPS, compass and related communications capabilities. UDIs are obviously highly technical and useful underwater tools used by military, search and rescue and exploration teams but are readily accessible to recreational divers needing underwater communications and boat connectivity.

10. **DRA** A similar development from Dan Europe (DSL) is the Diver Safety Guardian (DSG) software package providing the diver with feedback from an online Deco Risk Analyzer (DRA). Based on permissible gradients, it is under testing and development. An EOD risk estimator now, plans are in the works to make it a wet (OTF) risk estimator

11. **CCPlanner** CCPlanner is a LANL software package offering full RGBM, modified RGBM, USN M-value and Buhlmann Z-value algorithms for dive planning. It is used by the C&C Team and is not distributed commercially but is obtainable under written contract. Also encoded is the Hills TM. It is also provided with licensed LANL RGBM codes. A risk analysis routine using the LANL DB is encoded in CCPlanner and embedded in licensed RGBM OC and RB codes.

Output is typically extensive from modern diveware. Platforms range from PCs to Droid devices as well as Workstations to Mainframes. Languages employed in codes include VIZ, BASIC, FORTRAN, C, and derivatives. Meter Vendors (Suunto, Mares, Liquivision, UTC, Atomic Aquatics, Cressisub, Sherwood, Oceanic, Genesis, Shearwater, Uwatec, Cochrane and Aeris to name a few) often supply proprietary software packages keyed to their meter algorithms for coupled dive planning. Models are varied across GMs and BMs [8].

Examples

Sample output from 3 diveware packages are listed next. Example 1 is output from Abyss, Example 2 is output from GAP and Example 3 is output from CCPlanner. The profile entries are evident and the flow is easy to follow. In addition to decompression scheduling packages also track oxygen toxicity.

Output from all diveware is similar to the offerings listed here. Essential features are always decompression schedules, ascent and descent rates, ambient surface pressure, CNS and full body toxicity estimates, gas switches and CCR setpoints and user conservatism knob settings. The largest contingent of users are obviously technical divers, professional divers, scientific divers and exploration divers across military, recreational, research, training and commercial sectors.

As another look at diveware and its usefulness, consider the following comparison of open circuit and rebreather diving using standard diving software. Dive computers and coupled software packages provide great flexibility and offer the diver many choices for optimal staging and alternative strategies on OC and RB systems. The following example contrasts OC and RB diving for the same depth-time profile and reflects the ICD mitigation protocol discussed earlier.

Rebreather and Open Circuit Profiles

Rebreathers hold oxygen partial pressures, pp_{O_2}, or oxygen fractions, f_{O_2}, constant in the breathing mix. Closed circuit rebreathers (CCR) hold the oxygen partial pressure, pp_{O_2}, constant (setpoints). Open circuit (OC) regulators, of course, operate at constant breathing mix fractions of oxygen, helium and nitrogen, that is, f_{O_2}, f_{He} and f_{N_2}. Rebreathers have decompression advantages over open circuit devices because oxygen fractions or partial pressures can be held at higher levels forcing gas loadings of nitrogen and helium lower but not withstanding oxtox concerns. One way or another diving is constrained by DCS and/or OT.

ABYSS-2000, Advanced Dive Planning Software

This Short table printed for: Bruce Wienke on February 18, 2006

Profile: Abyss1 [Template=DEFAULT]

ATTENTION: The fact that this table was generated by ABYSS does not guarantee freedom from the possibility of decompression sickness. Diving is an inherently activity that may result in injury or death. Following this Abyss diving profile does not assure me that I won't be injured or killed. Decompression, Deep Diving, Ca Penetration and the use of Mixed Gas while diving are extremely hazardous aspects of an already dangerous activity.

Surface Altitude, 0 ft Safety Altitude, 985 ft Algorithm, Abyss 100

J-Factors: Depth = +0% Bottom Time = +0% N2 = +0% He = +0% Ne = +0% Ar = +0%

Depth (Ft)	Time at	Run Time	Gas & Percent O2%	N2%	He%	Status
0	0.0	0.0	16	54	30	Surface
240	20.0	22.6	16	54	30	Entered by user
70	2.0	30.2	16	54	30	DECO
60	3.0	33.6	16	54	30	DECO
50	5.0	38.9	16	54	30	DECO
40	8.0	47.2	16	54	30	DECO
30	13.0	60.6	16	54	30	DECO
20	24.0	84.9	16	54	30	DECO
10	61.0	146.2	16	54	30	DECO
0	0.3	0.0	16	54	30	Surface

DECO Stops Depth	Time	Run	Gas & Percent O2%	N2%	He%
70	2.0	30.2	16	54	30
60	3.0	33.6	16	54	30
50	5.0	38.9	16	54	30
40	8.0	47.2	16	54	30
30	13.0	60.6	16	54	30
20	24.0	84.9	16	54	30
10	61.0	146.2	16	54	30

DIVE SUMMARY

Run Time	146.6 Min	Deco Time	116.0 Min
CNS Clock	14.14%	OTU's	36.02
Max PPO2	1.32 (Atm)	Min PPO2	0.16 (Atm)
Max END	158.08 (Ft)	Max Workload	Mild
RMV	0.50 (CuFt)	Max Depth	240.00 (Ft)

Gas Consumption Gas & Percent O2%	N2%	He%	Volume (CuFt)	Reserve (CuFt)
16	54	30	257.0	0.0
			257.0	0.0

Total Gas Consumed	256.96
Required reserve	No reserve
Total Reserves	0.00
Total Gas Required	256.96 (CuFt)

Open circuit (OC) divers optimize their dive time while minimizing decompression requirements by making gas switches at various depths. Obviously, gas switching can be dialed to hold gas fractions constant on ascent, or we should say, relatively constant on ascent with constancy increasing with number of switches. The limit to the number of switches is logistic, of course, depending on depth, time and the capacity of divers to carry switch gases and/or tie switch gases off on an ascent line. The former limitation confronts technical divers operating without support teams able to string decompression tank lines. The latter limitation is one of cylinder availability. Or bodies. Not discounting logistics, it is interesting to compare rigging open circuit diving to mimic rebreather diving and decompression debt versus gas switching at constant pp_{O_2}.

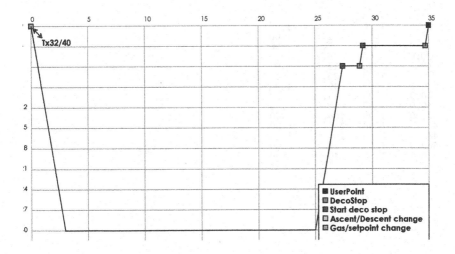

NAUI GAP - Dive Information

Created for: brwswe	Page 1/1	Creation Date 2/17/2006

The information presented here does in no way guarantee that you will not be injured or killed.
The authors accept no liability for your use of the information presented here.

deco schedule

NAUI-GAP (1530)

Model: RGBM   classic RGBM
mode: OC

Depth	Time (RT)	Gas	PO2
30 m	22.0 (25)	Tx32/40	1.28
6 m	2 (29)	Tx32/40	0.51
3 m	6 (35)	Tx32/40	0.42
0 m	– (35)	Tx32/40	0.00

Breathing gas	Volume
Tx32/40	2080.50 lit

Total CNS	Total OTU
13.5 %	35.8

Maximal Depth	Maximal Time
30.0 m	00:35 (= 35 min)

Maximal Po2 (B)	Maximal Po2 (D)
1.4 Bar	1.6 Bar

Maximal RMV (B)	Maximal RMV (D)
20.0 lit/min	10.0 lit/min

Maximal END	
30.0 m	

CCPlanner Tmx 16-34

C&C Operations -- OC
Version 6/20/14

  --Dive Sequence Number--   1

  --Scaling/Control Flags--

  temperature = 303 K   repet flag =   1
  dive number = 1 altitude =   0.0 ft
  rfac = 0.85   pfac = 1.00   bfac = 0.35
  unsat = 1
  surface consumption rate = 0.75 ft^3/min
  tissue/m-value sets jmix = 1 jndl = 1 nbub = 0
  phase constant =   840.0 microns^3
  zh = 0.61   zn = 0.81   trat = 1.03

  --Dive Profile--

  time since last dive = 1440. min
  time of last dive =   0. min
  ave depth last dive =   0. fsw

  surface air = 0.79 nitrogen 0.00 helium down switches = 2

   switch 1 depth =   0.0 fsw helium = 0.00 nitrogen = 0.79
              speed =   60.0 fsw/min   way time =   0.0 min
   switch 2 depth =   190.0 fsw helium = 0.34 nitrogen = 0.50
              speed =   -30.0 fsw/min   way time =   17.5 min

  bottom trimix = 0.34 helium 0.50 nitrogen up switches = 1

   switch 3 depth =   20.0 fsw helium = 0.00 nitrogen = 0.00
              speed =   -30.0 fsw/min   way time =   0.0 min

  fit parameters   rfit = 0.8589 bfit = 0.4697 pfit = 1.0000
                   rn = 0.9824   rh = 0.8596 r0 = 0.6718
                   depmx = 190.0 depav = 190.0

  ---Metrics And Debugging Parameters---

  rh = 0.8596 microns   rn = 0.9824 microns
  phase volume =   505. microns^3
  bfac/sfac/efac/tfac/dfac/gfac/rfac = 0.35/0.84/1.00/0.81/0.96/1.00/0.85/
  r0/rc/rstr/trat = 0.64/0.64/0.38/1.03/

tissue	factor	tissue	factor
1.0	1.00	0.3	1.00
2.0	1.00	0.7	1.00
5.0	1.00	1.7	1.00
10.0	1.00	3.3	1.00
20.0	1.00	6.7	1.00
40.0	1.00	13.3	1.00
80.0	1.00	26.7	1.00
120.0	1.00	40.0	1.00
160.0	1.00	53.3	1.00
240.0	1.00	80.0	1.00
320.0	1.00	106.7	1.00
400.0	1.00	133.3	1.00
480.0	1.00	160.0	1.00
560.0	1.00	186.7	1.00
720.0	1.00	240.0	1.00

```
trimix
fc = 1.00
alt =    0.  fn = 0.50  fh = 0.34  fo = 0.16
```

	nitrogen		helium		
depth	ndl	half	ndl	half	
40.	190.3	240.0	190.3	80.0	0.0
50.	103.8	160.0	103.8	53.3	0.0
60.	54.0	80.0	54.0	26.7	0.0
70.	43.0	80.0	43.0	26.7	0.0
80.	36.0	80.0	36.0	26.7	0.0
90.	21.3	40.0	21.3	13.3	0.0
100.	18.4	40.0	18.4	13.3	0.0
110.	11.4	20.0	11.4	6.7	0.0
120.	10.0	20.0	10.0	6.7	0.0
130.	9.0	20.0	9.0	6.7	0.0
140.	8.2	20.0	8.2	6.7	0.0
150.	7.5	20.0	7.5	6.7	0.0
160.	5.2	10.0	5.2	3.3	0.0
170.	4.8	10.0	4.7	3.3	0.0
180.	4.4	10.0	4.4	3.3	0.0
190.	4.1	10.0	4.1	3.3	0.0
200.	3.8	10.0	3.8	3.3	0.0

m-values/mixed values

	nitrogen			helium			mixed	
half	mO	dm	half	mO	dm	mxmO	mxdm	
1.0	119.8	2.12	0.3	153.3	2.59	133.4	2.31
2.0	115.0	2.04	0.7	147.4	2.48	128.1	2.22
5.0	100.5	1.79	1.7	129.8	2.17	112.4	1.94
10.0	81.7	1.47	3.3	103.5	1.70	90.5	1.56
20.0	68.3	1.27	6.7	83.8	1.40	74.6	1.32
40.0	60.2	1.19	13.3	72.0	1.28	65.0	1.22
80.0	53.9	1.12	26.7	63.0	1.18	57.6	1.14
120.0	51.1	1.10	40.0	59.3	1.14	54.4	1.12
160.0	49.4	1.08	53.3	57.3	1.12	52.6	1.10
240.0	47.1	1.06	80.0	55.5	1.11	50.5	1.08
320.0	45.8	1.05	106.7	55.0	1.10	49.5	1.07
400.0	44.9	1.05	133.3	54.7	1.09	48.9	1.07
480.0	44.2	1.04	160.0	54.5	1.09	48.4	1.06
560.0	43.7	1.04	186.7	54.4	1.08	48.0	1.05
720.0	42.7	1.02	240.0	54.1	1.07	47.3	1.04

tissue	tissue	radius	gradient
0.3	1.0	0.4503	103.2
0.7	2.0	0.4138	133.5
1.7	5.0	0.3828	163.9
3.3	10.0	0.3735	177.2
6.7	20.0	0.3681	184.6
13.3	40.0	0.3652	188.5
26.7	80.0	0.3636	190.5
40.0	120.0	0.3631	191.1
53.3	160.0	0.3628	191.5
80.0	240.0	0.3626	191.8
106.7	320.0	0.3624	192.0
133.3	400.0	0.3623	192.1
160.0	480.0	0.3623	192.2
186.7	560.0	0.3622	192.2
240.0	720.0	0.3622	192.3

CCPlanner Tmx 16-34

--Decompression Schedule--

iters = 4 staging algo = 0 dcut = 80.0 fsw
bottom depth = 190.0 fsw ppo2 = 1.1 atm OTU/CNS = 19.8 min/0.06 %
bottom time = 20.7 min

depth (fsw)	wait (min)	tissue (min)	tension (fsw)	pss (fsw)	ppo2 (atm)	OTU (min)	CNS (%)	gas (ft^3)
190.0 -	17.5			-	1.1 -	19.8 - 0.06 -		98.
130.0 -	0.0 -	3.7 -	144.6 -	31.6 -	0.8 -	0.0 - 0.00 -		1.
120.0 -	0.0 -	3.7 -	141.4 -	31.6 -	0.7 -	0.0 - 0.00 -		1.
110.0 -	1.0 -	3.7 -	128.5 -	31.5 -	0.7 -	0.5 - 0.00 -		4.
100.0 -	1.0 -	3.7 -	121.2 -	31.5 -	0.6 -	0.4 - 0.00 -		2.
90.0 -	2.0 -	7.3 -	109.8 -	31.6 -	0.6 -	0.5 - 0.00 -		6.
80.0 -	2.0 -	7.3 -	99.7 -	31.6 -	0.5 -	0.3 - 0.00 -		6.
70.0 -	2.0 -	7.3 -	90.5 -	31.5 -	0.5 -	0.0 - 0.00 -		5.
60.0 -	3.0 -	14.6 -	80.9 -	31.2 -	0.5 -	0.0 - 0.00 -		7.
50.0 -	4.5 -	14.6 -	71.1 -	31.1 -	0.4 -	0.0 - 0.00 -		9.
40.0 -	5.5 -	14.6 -	60.6 -	31.0 -	0.4 -	0.0 - 0.00 -		10.
30.0 -	8.5 -	29.2 -	50.2 -	30.4 -	0.3 -	0.0 - 0.00 -		13.
20.0 -	4.5 -	29.2 -	39.6 -	30.2 -	1.6 -	8.7 - 0.07 -		6.
10.0 -	7.5 -	58.4 -	29.2 -	29.3 -	1.3 -	11.1 - 0.05 -		7.
	68.5					41.2	0.19	176.

deco plus surfacing time = 47.8 min
cum CNS% = 0.19 cum OTU = 41.2 min
cum gas consumption = 176. ft^3

dive time = 68.5 min   excitation depth = 190.0 fsw

total dives processed = 1

Let's compare a test dive to $250\,fsw$ for $15\,min$, using a CCR with setpoint of $pp_{O_2} = 1.2\,atm$ and a set of doubles (triples) filled with TMX 14/56. The diluent is also taken to be TMX 14/56. At $250\,fsw$, $pp_{O_2} = 1.2\,atm$, and $pp_{N_2} = 85\,fsw = 2.6\,atm$ for both OC and RB divers. With that said, we first compare raw RGBM decompression profiles for the exposure, that is, no gas switches on OC. This gives the CCR baseline too. Descent and ascent rates are 90 and $30\,fsw/min$, respectively. Both central nervous system (CNS) and full body (pulmonary) toxicity are not concerns here, just minimization of decompression time. The comparative decompression schedules are given in Table 1. The first two columns contrast CCR and no switch OC staging. These are nominal RGBM profiles consistent with released tables and licensed software.

Next, let's make a switch on OC at $125\,fsw$, the midpoint of the dive, to TMX 25/45. Here, $pp_{O_2} = 1.2\,atm$, as required, while $pp_{N_2} = 47.5\,fsw = 1.44\,atm$. Columns 1 and 3 tabulate the CCR and OC decompression schedules for quick comparison. The single switch on OC reduces decompression time by roughly 35% compared to the previous schedule.

Finally, let's make gas switches on OC at 200, 150, 100 and $50\,fsw$ to TMX 17/53, 22/48, 30/40 and 48/22 respectively. Note that the nitrogen fraction stays constant, that is, $f_{N_2} = 0.30$, as we increase oxygen and decrease helium keeping $pp_{O_2} = 1.2\,atm$ as discussed previously in ICD mitigation. Columns 1 and 4 now

Table 1 Comparative rebreather and open circuit profiles

| Depth (fsw) | Rebreather | | Open Circuit | | | | | |
| | Stop (min) | pp_{O_2} (atm) | No switches | | 1 switch | | 4 switches | |
			Stop (min)	pp_{O_2} (atm)	Stop (min)	pp_{O_2} (atm)	Stop (min)	pp_{O_2} (atm)
250	15	1.2	15	1.2	15	1.2	15	1.2
160	0.0	1.2	0.5	0.9	0.5	0.9	0.5	0.9
150	0.0	1.2	1.0	0.8	1.0	0.9	0.5	1.2
140	0.5	1.2	1.0	0.8	1.0	0.8	0.5	1.2
130	0.5	1.2	1.0	0.7	1.0	0.7	1.0	1.1
120	0.5	1.2	1.5	0.6	1.0	1.2	1.0	1.0
110	0.5	1.2	1.5	0.6	1.0	1.1	1.0	1.0
100	1.0	1.2	2.5	0.6	2.0	1.0	1.5	1.2
90	1.0	1.2	3.0	0.6	2.0	0.9	1.5	1.1
80	1.5	1.2	3.5	0.5	2.0	0.9	2.0	1.0
70	1.5	1.2	3.5	0.4	2.5	0.8	2.0	0.9
60	2.0	1.2	6.5	0.4	3.5	0.7	2.5	0.8
50	2.0	1.2	8.0	0.4	5.5	0.6	3.5	1.2
40	3.5	1.2	8.5	0.3	6.5	0.6	4.0	1.1
30	3.5	1.2	15.5	0.3	7.0	0.5	4.5	0.9
20	4.0	1.2	19.5	0.2	15.5	0.4	8.5	0.8
10	7.5	1.2	33.0	0.2	20.5	0.3	12.5	0.6
	55.6		136.1		98.6		73.1	
	OTU=58.8 (min)		OTU=23.4 (min)		OTU=36.4 (min)		OTU=56.1 (min)	
	CNS=0.22		CNS=0.09		CNS=0.13		CNS=0.18	

clock CCR and OC decompression schedules with the 4 gas switches indicated. Further reduction in decompression time on OC is clearly evident. Decompression time has been cut in half, only $20\,min$ longer than the CCR decompression time. Note that full body (OTU) and central nervous system CNS clock estimates for OC fall below CCR values as expected.

CCPlanner was used for compiling Table 1 and the RB dive resides in the LANL DB. The OC dives are hypothetical for illustrative purposes.

Diveware Model Characteristics Comparison

Of course there are many models that fall into GM and BM categories and some are depicted in Fig. 2 of chapter "Computer Profile Data" which contrasts model staging predictions for the USN deep stop air tests at $170\,fsw$ for $30\,min$. Roughly one might say that the models with long time tails in the shallow zone are GMs while those with more rapid ascent and shorter time tails in the shallow zone are

BMs. Depending on user settings, dive computers and diveware can mimic all sets of model staging options easily. Questions of right or wrong then relate to model correlations with diving data and that is where the correlated and tested USN, ZHL, VPM and RGBM algorithms weigh more heavily. For reliability and reproducibility that is also important. Diveware is a valuable and quick tool for comparing different staging options. A risk comparison for the USN deep stop tests is summarized in Table 2 against the ZHL and VPM. Results are generic for GM and BM dive computers and dive planning software.

Comparative GM and BM Risk Estimates for Wet Pod Trials

To compare GM and BM risk estimates with actual data, the recent NEDU deep stop test suffices. Some 100+ air dives to 170 fsw for 30 min were staged in a wet pod at NEDU with a DCS incidence rate of $r = 5.5\%$. The staging model used was called the bubble volume model (BVM3) keyed to dissolved gas volume content not bubbles and is a USN model. The decompression schedule is listed in Table 2 along with the corresponding VPM schedule. The VPM profile risk is denoted r_{VPM}. The ZHL profile is listed in the last column with risk r_{ZHL} and is very similar to the standard USN Tables. The VPM and ZHL risks are also very close in this case with run times about the same too. That is interesting too.

	Depth	NEDU time	VPM time	ZHL time
Table 2 Comparative NEDU deep stop air schedules and risk	(fsw) (*min*)	(*min*)	(*min*)	(*min*)
	170	30.0	30.0	30.0
	100		1.0	
	90		2.0	
	80		2.5	
	70	12.0	3.0	
	60	17.0	4.5	
	50	15.0	5.5	1.5
	40	18.0	7.5.	6.0
	30	23.0	11.0	9.0
	20	17.0	15.0	17.5
	10	72.0	23.0	44.0
		206.0	114.5	116.5
	$r_{VPM} = 8.1\%$	$r_{VPM} = 2.3\%$	$r_{USN} = 2.1\%$

The EOD risks were computed with CCPlanner, a LANL software package, with encoded USN and VPM model implementations. The VPM risk prediction is larger than the reported incidence rate, $r = 5.5\%$, thus suggesting the EOD fit is conservative. The OTF risk estimates in Table 1 also slightly overestimate the USN 1% risk limits for nonstop air diving and similarly are conservative.

References

1. Workman RD, *Calculation Of Decompression Schedules For Nitrogen-Oxygen And Helium-Oxygen Dives*, 1965; USN Experimenal Diving Unit Report, NEDU, 6–65, Washington DC.
2. Buhlmann AA, *Decompression: Decompression Sickness*, Springer-Verlag Publishing, 1984; Berlin.
3. Yount DE and Hoffman DC, *On The Use Of A Bubble Formation Model To Calculate Diving Tables*, Aviat. Space Environ. Med. 1986; 36: 149–156.
4. Wienke BR, *Reduced Gradient Bubble Model*, Int. J. Biomed. Comp. 1990; 26: 237–246.
5. Wienke BR and O'Leary TR, *Diving Bubble Model Data Correlations*, J. Marine Sci. Res. Dev. 2016; 6(4): 1000204.
6. Wienke BR, *Biophysics And Diving Decompression Phenomenology*, Bentham Science Publishers, 2016; Sharjah.
7. Wienke BR, *Deep Stop Model Correlations*, J. Bioeng. Biomed. Sci. 2015; 5: 12–18.
8. Wienke BR, *Science of Diving*, CRC Press, 2015, Boca Raton.

Bubble Issues and Dive Computer Implementations

Bubble birth, growth, evolution, destruction and elimination in the body of human divers are central issues in safe diver staging protocols from exposures at depth. Today, despite incredible technological advances in medical and physiological science, we really know very little about bubbles in vivo and their complex behavior under pressure and environmental changes. Measuring bubbles and their properties in vivo by invasive means often destroys or changes what is being measured. Measuring with non-invasive techniques is very limited. Doppler scoring of moving body bubbles is only able to count numbers. Experiments using materials with properties similar to blood and tissue can be useful as a starting point for simulating bubble behavior but of course blood and tissue are metabolic and perfused adding additional complexity and unknowns to coupled modeling and simulation. In this vein, therefore, we investigate bubble data and experiments in the laboratory to make some hypothetical estimates of possible impacts of bubble regeneration and broadening on diver staging regimens. We emphasize these are speculative based on experiments in the laboratory not in the field nor in divers. The fact that tissue and blood are both perfused and metabolic always throws wrenches into biophysical modeling. Simple rheological assignments of laboratory variables and physical constants doesn't necessarily extrapolate to divers. Dive computer model [1–6] implementations await additional information about bubble birth, growth, evolution, destruction and elimination of bubbles in divers.

Bubble Dynamics

Most questions of seed distributions, lifetimes, persistence and origins in the body are unanswered today. And while we have yet to measure microbubble distributions and lifetimes in the body, we can gain some insight from laboratory measurements

B. R. Wienke, T. R. O'Leary, *Understanding Modern Dive Computers and Operation*, SpringerBriefs in Computer Science, https://doi.org/10.1007/978-3-319-94054-0_9

and statistical mechanics. Microbubble distributions have been studies extensively. The biophysics work [7–10, 13] details some interesting studies about microbubbles and properties in general and follows in abbreviated form.

Microbubbles typically exhibit size distributions that decrease exponentially in radius, r. Holography measurements of cavitation nuclei in water tunnels suggest [12],

$$N = N_0 \, exp \, (-\beta r)$$

with,

$$N_0 = 1.017 \times 10^{12} \, m^{-3}$$

$$\beta = 0.0512 \, micron^{-1}$$

Experiments in gels also display exponential dependencies in cavitation radii,

$$N = N_0 \, exp \, (-\alpha r)$$

with,

$$N_0 = 662.5 \, ml^{-1}$$

$$\alpha = 0.0237 \, micron^{-1}$$

Both MRI and Doppler laser measurements of water and ice droplets [11] in the atmosphere underline exponential decrease in number density as droplet diameter increases. Ice and water droplets in clouds typically range, $2 \, micron \, \leq \, r \, \leq \, 100 \, micron$. Dust and pollutants are also exponentially distributed, potentially serving as heterogeneous nucleation sites. It might be a surprise if micronuclei in the body were not exponentially distributed in number density versus size.

Lifetimes of cavitation voids are not known, nor measured, in the body. The radial growth equations provide a framework for estimation using nominal blood and tissue constants. Consider first the mass transfer equation,

$$\frac{\partial r}{\partial t} = \frac{DS}{r} \left[\Pi - P - \frac{2\gamma}{r} \right]$$

with all quantities as before, that is, r bubble radius, D diffusivity, S solubility, γ surface tension, P ambient pressure and Π total gas tension. The time to collapse, τ, can be obtained by integrating over time and radius, taking initial bubble radius, r_i,

$$\tau = \int_0^\tau dt = \int_{r_i}^0 \left[\frac{r}{DS}\right]\left[\frac{1}{\Pi - P - 2\gamma/r}\right] dr$$

$$= \left[\frac{\Delta p r_i (4\gamma + \Delta p r_i) - 8\gamma^2 \ln (1 - \Delta p r_i/2\gamma)}{2DS\Delta p^3}\right]$$

with,

$$\Delta p = P - \Pi$$

If surface tension is suppressed, we get,

$$\tau = \frac{r_i^2}{2DS\Delta p}$$

In both cases, small tension gradients, Δp and small transport coefficients, DS, lead to long collapse times and vice-versa. Large bubbles take a longer time to dissolve than small bubbles. Taking nominal transport coefficient for nitrogen, $DS = 56.9 \times 10^{-6}$ $micron^2/sec\,fsw$, and initial bubble radius, $r_i = 10.0\,micron$, for $\Delta p = 3.0\,fsw$ and $\gamma = 40\,dynes/cm$, we find,

$$\tau = 0.25\,sec$$

In the Rayleigh-Plesset picture [11], the radial growth equation takes the form, neglecting viscosity,

$$\left[\frac{\partial r}{\partial t}\right]^2 = \frac{2(\Pi - P)}{3\rho}\left[\frac{r_i^3}{r^3} - 1\right] + \frac{2\gamma}{\rho r}\left[\frac{r_i^2}{r^2} - 1\right]$$

so that the collapse time by diffusion only is,

$$\tau = \int_0^\tau dt = \left[\frac{3\rho}{2(\Pi - P)}\right]^{1/2} \int_{r_i}^0 \left[\frac{r_i^3}{r^3} - 1\right]^{-1/2}$$

$$dr = r_i \frac{\Gamma(5/6)}{\Gamma(1/3)}\left[\frac{3\pi\rho}{2\Delta p}\right]^{1/2}$$

with,

$$\Gamma(5/6) = 1.128$$

$$\Gamma(1/3) = 2.679$$

Suppressing the diffusion term in the growth equation, there similarly obtains,

$$\tau = \int_0^\tau dt = \left[\frac{\rho}{2\gamma}\right]^{1/2} \int_{r_i}^0 r^{1/2} \left[\frac{r_i^2}{r^2} - 1\right]^{-1/2}$$

$$dr = r_i \frac{\Gamma(-3/4)}{\Gamma(-1/4)} \left[\frac{\pi \rho r_i}{4\gamma}\right]^{1/2}$$

with,

$$\Gamma(-3/4) = -4.834$$

$$\Gamma(-1/4) = -4.062$$

Collapse time in the Rayleigh-Plesset picture is linear in initial bubble radius, r_i and inversely proportional to the square root of the tension gradient, Δp, or the surface tension, γ. Taking all quantities as previously, with density, $\rho = 1.15\, g/cm^3$, we find with surface tension suppressed,

$$\tau = 2.91 \times 10^{-3}\, sec$$

and, for the diffusion term suppressed with only the surface tension term contributing,

$$\tau = 2.52 \times 10^{-6}\, sec$$

Dissolution times above range,

$$10^{-6}\, sec \le \tau \le 10^{-1}\, sec$$

In the Yount model of persistent nuclei, within the permeable gas transfer region, seed nuclei lifetimes, τ, range,

$$10^{-6}\, sec \le \tau \le 10^{-2}\, sec$$

The collapse rate increases with both γ and Δp and inversely with r_i. Small bubbles collapse more rapidly than large bubbles, with large bubble collapse driven most by outgassing diffusion gradients and small bubble collapse driven most by constrictive surface tension. Between these extrema, both diffusion and surface tension play a role. In any media, if stabilizing material attaches to micronuclei, the effective surface tension can be reduced considerably and bubble collapse arrested temporarily, that is, as $\gamma \to 0$ as a limit point. For small bubbles, this seems more plausible than for large bubbles because smaller amounts of material need adhere. For large bubbles, bubble collapse is not aided by surface tension as much as for small bubbles, with outgassing gradients taking longer to dissolve large bubbles than small ones. In both cases, collapse times are likely to lengthen over the short times estimated above. Additionally, external influences on the bubble, like crevices and surface discontinuities, may prevent bubble growth or collapse. All this adds to bubble complexities faced by modelers. The questions of regeneration and broadening are equally complex and follow.

Bubble Regeneration

It is known that samples of knox gelatin, fingerling salmon and albino rats [7–9] display increased resistance to bubble formation following rapid application of bubble crushing pressures. The larger the compression pressures, the fewer bubbles that form with the same allowed supersaturation during decompression. Bubble models assume that gas nuclei are *crushed* by the mechanical strength of an initial compression and that the number of nuclei larger than the critical radius, ϵ, decreases. Presumably, surfactant molecules are forced out of the bubble skin into a possible reservoir outside where they remain available to retake their old positions in the bubble distributions. This is also the reason to make first dives deepest and subsequent dives shallower than the previous. While this has not been proven nor tested in the diver per se, it forms a plausible background to take a look at bubble regeneration effects in diver decompression staging.

It has also been shown theoretically [10] at equilibrium that the radial distribution of bubble nuclei is exponential and that a nuclear population, once crushed, stochastically restores itself over time scales of minutes to hours to days. This effect is seen everyday in reactor coolants, bubble column processes, high speed flows and high temperature chemical reactions. The higher the flow rates and temperatures, the shorter the regeneration times. In boiling water reactors for instance, regeneration times are in the 10 sec range. Stationary decompressed gels in the laboratory exhibit long regeneration times of days. Bubble regeneration over varying time and size scales is seen in surf and breaking waves [11]. It seems that rapidly moving, turbulent and high temperature environments more readily support regeneration because of increased collisional dynamics and material disruption. The diver by comparison seems an unlikely candidate but perhaps tribonuncleation processes in tissue and blood [13] are strong enough to initiate bubble regeneration too. Nothing is presently known about possible regeneration times in humans and divers and the assumption has been that regeneration times are long compared to dive times. We will look at bubble regeneration effects on diver staging protocols for both short and long times. The time span for bubble regeneration in bubble models is a crucial element in calculations of NDLs and decompression staging.

Using empirical laboratory data, estimates of hypothetical bubble regeneration effects have been collated by the Authors [19] and codified for use in dive computers and diveware within the USN, ZHL, VPM and RGBM frameworks. Hypothetical regeneration time scales less than a day or so impact diver staging regimens. Hopefully such analyses are of interest to dive computer vendors, diveware purveyors and table fabricators. Some day issues of regeneration will be resolved by direct measurements in the body.

Bubble Broadening

Bubble broadening (Ostwald ripening) is a phenomena observed by Ostwald [14] in 1897 whereby small bubbles diminish in size and large bubbles grow over time spans of hours to days. Concentration gradients (diffusion) drive the transport of material across bubble interfaces with small bubbles at higher concentrations than large bubbles because of their increased curvature and surface tension pressure. An everyday example is the re-crystalization of water within ice cream which gives old ice cream a gritty, crunchy texture. Larger ice crystals grow at the expense of smaller ones within the ice cream creating a coarser surface texture.

A systematic theory of bubble broadening developed by Lifshitz, Slyozov and Wagner [15] (LSW) suggests in supersaturated and solid solutions that the distribution mean bubble radius, r_m, evolves in time as,

$$r_m^3 = r_0^3 + Kt$$

with r_0 the unbroadened initial mean radius and K the transport coefficient a function of temperature, bubble surface tension, diffusivity, gas solubility and gas molar volume. For a wide range of experiments [16] the relationship holds with the transport coefficient, K, varying across materials of course. A couple experiments impacting possible dive computer and diveware implementations of bubble broadening follow.

In the Kabalnov [17] fluorocarbon experiments, the LSW transport coefficient, K, was determined,

$$K_{fluor} = 5.2 \times 10^3 m^3 sec^{-1}$$

for the emulsion. Rheological scaling suggests the extrapolation to body blood and tissue,

$$K_{blood} = \frac{K_{fluor}}{5.7}$$

with,

$$r_0 = 16.56 \, micron$$

In the Del Cima [18] glycerol-water (75–25%) studies, it was found that $K_{glycerol}$ deviated from the LSW value,

$$K = 6.1 \times 10^3 m^3 sec^{-1}$$

with

$$r_0 = 18.42 \, micron$$

according to,

$$K_{glycerol} = 2.0 \times 10^7 m^{1/0.1956} sec^{-1}$$

with measured mean radius, r_m, in time t, fitted by,

$$r_m = [16.977^3 + 14203.0 \, t^{0.67637}]^{1/3}$$

for t in hrs and r_m in $micron$ ($10^{-6} m$). Glycerol-water solutions of course are not blood and the transport coefficient in glycerol-water is roughly 1/8 the value in blood. Using the fitted expression one then takes for tissue and blood,

$$K_{glycerol} = \frac{K_{blood}}{8}$$

as an approximation. Other representations in different materials with rheological scalings can be found in the literature.

As with bubble regeneration using empirical laboratory data, estimates of hypothetical bubble broadening effects have been collated by the Authors [19] and codified for use in dive computers and diveware within the USN, ZHL, VPM and RGBM frameworks [1–6]. Hypothetical broadening time scales less than $2 \, hrs$ or so impact diver staging regimens. Beyond $2 \, hrs$ broadening effects are negligible. Such analyses are of interest to dive computer vendors, diveware purveyors and table fabricators. As above, hopefully issues of broadening will be resolved by direct measurements in the body one day.

References

1. Workman RD, *Calculation Of Decompression Schedules For Nitrogen-Oxygen And Helium-Oxygen Dives*, 1965; USN Experimenal Diving Unit Report, NEDU, 6–65, Washington DC.
2. Buhlmann AA, *Decompression: Decompression Sickness*, Springer-Verlag Publishing, 1984; Berlin.
3. Yount DE and Hoffman DC, *On The Use Of A Bubble Formation Model To Calculate Diving Tables*, Aviat. Space Environ. Med. 1986; 36: 149–156.
4. Wienke BR, *Reduced Gradient Bubble Model*, Int. J. Biomed. Comp. 1990; 26: 237–246.
5. Wienke BR and O'Leary TR, *Diving Bubble Model Data Correlations*, J. Marine Sci. Res. Dev. 2016; 6(4): 1000204.
6. Wienke BR, *Dive Computer Profile Data And On The Fly And End Of Dive Risk Estimators*, J. Appl. Biotech. Bioeng. 2018; 5(1): 00118.

7. Yount DE and Strauss RH, *Bubble Formation In Gelatin: A Model For Decompression Sickness*, J. Appl. Phys. 1976; 47: 5081–5089
8. Yount DE, Gillary EW, and Hoffman DC, *A Microscopic Investigation Of Bubble Formation Nuclei*, J. Acoust. Soc. Am. 1984; 76: 1511–1521.
9. Yount DE, Yeung CM, and Ingle FW, *Determination Of The Radii Of Gas Cavitation Nuclei By Filtering Gelatin*, J. Acoust. Soc. Am. 1979; 65; 1440–1450.
10. Yount DE, *On The Evolution, Generation And Regeneration Of Gas Cavitation Nuclei*, J. Acoust. Soc. Am. 1982; 71: 1473–1481.
11. Wienke BR, *Science of Diving*, CRC Press, 2015, Boca Raton.
12. Mulhearn PJ, *Distribution Of Microbubbles In Coastal Waters*. J. Geophys. Res. 1981; 86: 6429–6434.
13. Wienke BR, *Biophysics And Diving Decompression Phenomenology*, Bentham Science Publishers, 2016; Sharjah.
14. Ostwald W, *Studien Uber Die Bildung Und Umwandlung Fester Korper*, Z. Phys. Chem. 1897; 22: 289–304.
15. Lifshitz IM and Slyozov VV, *The Kinetics Of Precipitation From Superheated Solid Solutions*, J. Phys. Chem Solids 1961; 19: 34–45; Wagner C *Theorie Der Alterung Von Neiderschlagen Durch Umlosen*, Z. Electrochemie 1961; 95: 581–597.
16. Baldan A, *Progress In Ostwald Ripening Theories And Their Applications To Nickel Based Superalloys*, J. Mat. Sci. 2002; 37: 2172–2202.
17. Kabalnov AS and Schukin ED, *Ostwald Ripening Theory: Application To Fluorocarbon Emulsion Stability*, J. Coll. Inter. Sci. 1992; 38: 69–97.
18. Del Cima OM, Oliveira PC, Rocha CM, Silva HS and Teixeira NC, *Gas Diffusion Among Bubbles And The DCS Risk*, Fluid Dyn. 2017; arXiv:1711.08987v1.
19. Wienke BR and O'Leary TR, On Bubble Regeneration And Broadening With Implications For Decompression Protocols, to be published Comp. Biol. Med.

Recap

Dive computers, protocols, models, data, risk and associated mathematical relationships were the topical focus. Basic principles were presented and then practical applications and results were detailed. Topics were related to diving protocols and operational procedures. Bubble dynamics and hypothetical issues such as bubble regeneration and broadening were described within model frameworks for possible implementation in dive computers, diveware and tables. The exposition linked phase mechanics to decompression theory with equations for computer implementation. Computer applications to mixed gas, decompression, open circuit (OC) and rebreather (RB) diving were described with correlated GM and BM software for comparisons and risk analyses. End of dive (EOD) risk estimators were statistically coupled to surfacing on the fly (OTF) risk estimators to extract OTF parameters using NLLS techniques across some 3569 profiles in the LANL DB. Supercomputers at LANL easily effected the calculations and correlations. Applications focused mainly on deep diving where risks increase and statistical collections and tabulations of data are very important. Dive computers of the future will build on present generation of models and algorithms to offer divers the most advanced hardware and software technology for the safest and most efficient ways to dive. As in all walks, computers are mainstays for progress, technological facility, operational ease and ranging applications. The biophysics of diving and decompression in the human body is extremely complex. More needs be learned to safely and routinely stage divers. Complexity links to both pressure and pressure changes and the reaction of the human body to same.

The physics, biology, engineering, physiology, medicine and chemistry of diving center on pressure and pressure changes. The average individual is subject to atmospheric pressure swings of 3% at sea level, as much as 20% a mile in elevation, more at higher altitudes and all usually over time spans of hours to days. Divers and their equipment can experience compressions and decompressions orders of magnitude greater and within considerably shorter time scales. While effects of pressure change are readily quantified in physics, chemistry and engineering applications,

B. R. Wienke, T. R. O'Leary, *Understanding Modern Dive Computers and Operation*, SpringerBriefs in Computer Science, https://doi.org/10.1007/978-3-319-94054-0_10

the physiology, medicine and biology of pressure changes in living systems are much more complicated. Caution is needed in transposing diving principles from one pressure range to another. Incomplete knowledge and biophysical complexities often prevent extensions of even simple causal relationships in biological science.

Gas exchange, oxygen toxicity, bubble formation and elimination and compression-decompression in blood and tissues in diving are governed by many factors, such as diffusion, perfusion, phase separation and equilibration, nucleation and cavitation, local fluid shifts and combinations thereof. Owing to the complexity of biological systems, multiplicity of tissues and media, diversity of interfaces and boundary conditions and plethora of bubble impacting physical and chemical mechanisms, it is difficult to solve the decompression problem in vivo. And equally difficult and elusive are direct measurements of bubbles, bubble sites and effective transport properties of tissues and blood in living human systems. Early decompression studies adopted the medical supersaturation viewpoint. Closer looks at the physics of phase separation and bubbles in the mid-1970s, and insights into gas transfer mechanisms, culminated in extended kinetics and dissolved-free phase theories embedded in today's computers. Expect new updated models and correlations will make their way into dive computers just as models discussed. Dive computers will be at the forefront of operational diving and technology improvements. And to coin a phrase from Bill Hamilton, "what works, works" in the dive protocol and dive computer business. What doesn't goes away fast. Amen.

Here are a few sidelights to the material in the form of answers to emails that have crossed our desks. Hope they are illuminating in both content and reply.

- *M-values and Z-values are limit points for bubble formation in divers?* Naw. In the early days, it was thought that M-values and Z-values were metastable dissolved gas limit points for spontaneous bubble formation (called *de novobubbles*). M-values and Z-values are neither limit points for de novo bubble formation nor any other mechanism. They are dissolved gas (statistical) limits for possible DCS symptomology in divers across arbitrary tissue compartments. They were formulated using DCS outcomes or Doppler scores against computed (GM) dissolved gas tensions by the medical community. Most M-values and Z-values for nonstop and light decompression diving have a DCS incidence rate below 1reported by the medical community. Remember most importantly, bubbles form on every dive.
- *Nucleation and cavitation are the same thing?* Not quite. The processes are distinctly different mechanisms. Unfortunately, terms are used interchangeably at times. Cavitation refers to the separation of liquid layers, holes, and defects and/or subsequent filling of voids with gas or vapor. Nucleation underscores the existence of preformed, very small, gas microemboli that possibly receive cavitation void gas or vapor and grow depending on conditions. The lifetimes of the gas microemboli are mostly unknown in divers. Gas nuclei are also thought to possibly stabilize for longer periods of time, seeding bubble growth after pressure changes independent of any cavitation. A debated issue in divers is the question of persistent micronuclei. And time scales for persistence.

- *Bubbles obey Boyle's Law under compression-decompression?* Nada again. Bubble shells in the body are normally lipid or aqueous substances. Under pressure and/or temperature changes, these substances render the gas inside as nonideal. Nonideal gases do not obey Boyle's Law though the departure is small for thin aqueous shells and more pronounced for thick lipid shells.
- *Helium NDLs are greater than nitrogen NDLs because nitrogen outgases slower?* No categorically until you reach the $200\, fsw$ level. At that point they are both very short. Otherwise, helium NDLs are shorter, just the opposite.
- *Does the product of gas diffusivity times solubility scale NDLs and decompression staging?* Yes actually, but you will not see this in available tables, software nor decompression meters popularly used these days. That may change.
- *Helium bubbles are smaller but more numerous than nitrogen bubbles in divers?* Not really known in divers, but seen in substrate experiments. Suggested by molecular properties of both gases. Computer bubble algorithms focusing on helium and nitrogen number distributions usually take the total bubble volume of both to be nearly the same in diver staging calculations. The cumulative volume of all bubbles in a bubble number distribution is called the phase volume and is used to limit diver ascents in BMs.
- *The high solubility of nitrogen generally makes nitrogen less desirable for diving?* For sure, especially when coupled to its molecular weight. But air is plentiful and cheap. And helium is expensive. Technical and professional divers do not dive *deep air* these days with something in the $130\, fsw$ range the limiting depth for air diving. Nitrox and helitrox are even better shallow gas choices than air.
- *Staging with trimix, it is advantageous to increase oxygen fraction with corresponding decrease in helium fraction on OC gas switches?* Yes, thereby keeping the nitrogen fraction relatively constant on ascent and decompression. Bottom mix should keep the nitrogen fraction as low possible and the oxygen fraction constrained to avoid oxygen toxicity. Rest is helium.
- *The Pyle stop (ad hoc) protocols correlate with the VPM and RGBM?* No, not rigorously but like the Bennett and Marroni prescriptions, stop protocols mimic broad features of bubble models.
- *The deep stop VPM and RGBM models can be tweaked to yield shallow stops?* Only using very strange bubbles or very large permissible bubble-dissolved gas gradients as occur with non stop diving in the shallow zones. Deep stop and shallow stop models converge for non stop diving in the recreational zones (less than $130\, fsw$).
- *The RGBM is patterned on Doppler bubbles?* No. Doppler bubbles are moving bubbles in the bloodstream coming from sites all over the body They correlate weakly with DCS incidence in divers excepting limb bends. The RGBM is correlated with DCS outcomes in actual profile data downloaded from dive computers.
- *RGBM nonstop times are shorter than Haldane nonstop times?* Mostly no, depending on mix. RGBM single dive nonstop limits and Haldane nonstop time limits are roughly the same across recreational diving and even technical diving.

- *The RGBM and VPM are correlated with diving data?* Definitely yes as stated over and over.
- *Difference between Haldane models and RGBM are not important for recreational diving?* Not quite. For single nonstop dives on air and nitrox, this is true. Not generally true for repetitive, multiday and deeper-than-first diving, with differences notable for short time intervals, large depth decrements between successive dives and deeper excursion for longer bottom times.
- *Technical RGBM and Recreational RGBM are different models?* No, they are the same bubble model in both cases. The Recreational RGBM has been streamlined for easy and fast dive computer implementation using M-values consistent with and obtained from the full Technical RGBM.
- *NAUI Recreational and Technical Tables have been tested in the field?* For sure and correlated with the RGBM Data Bank. See References for full descriptions of field testing and formal correlation with LANL Data Bank profile entries. Field testing is an ongoing process. The 100,000s of dives performed by NAUI recreational and technical divers (students and instructors) strongly attest to this fact. DCS spikes in RGBM diving are nonexistent.
- *Dive computers and diveware extrapolate outside nominal diving envelopes for the mixture and device (OC and RB) employed?* They extrapolate to any diving activity inside or outside nominal envelopes or they may shut down if depth exceeds programmed algorithm limitations. The User Manual often specifies algorithmic ranges of depth and altitude for which data correlations exist. Diving beyond the envelope is risky business.
- *Some dive computers offer both GM and BM algorithms as diver choice?* Yes, as seen earlier. For increasing depth and time, differences in staging regimens increase obviously and can be easily seen. GM and BM choices are best made by seasoned technical and professional divers using correlated algorithms or time tested protocols.
- *Aggressive-to-conservative knobs on computers and diveware vary widely across units and packages?* Definitely yes. And user changes in settings can produce large differences in staging options. Some 5% to 10% changes in critical parameters like M-values, Z-values, bubble radii, Boyle expansion factors, allowable surfacing bubble volume and others are the usual diver knobs and need to be plied carefully on the aggressive side. Nothing wrong diving with the most conservative settings?
- *Bubble models (BM) and dissolved gas models (GM) have the same NDLs for nominal settings?* Roughly true for air and nitrox dive computers using the USN, ZHL, VPM and RGBM algorithms and as can also be seen for the above set of computers and diveware. BM algorithms collapse to GM algorithms in the limit of small phase separation, usually the case for nonstop diving. For mixed gases like trimix and heliox there are wider variations in NDLs. Of course, for very deep diving NDLs approach zero and decompression is always requisite.
- *Computer models have been basically validated?* Yes, but a touchy question and depends on your viewpoint of *validation*. Certainly USN, ZHL, VPM and RGBM are considered validated from most user perspectives having been

computer dived safely for many years without DCS and oxtox spikes and an incidence rate below 1% roughly. The same is probably true of any computer model that has seen similar safe performance over time. On a more scientific level, USN, ZHL, VPM and RGBM have all been correlated with data of low DCS prevalence. The forwgoing text contains details of correlations. Laboratory and wet tests of models are selective and usually center on one or another profiles and not the full spectrum of diving across mixed gases, OC and RB systems deep and shallow exposures to name a few. Single tests on a single profile beg the question for some. That is where Data Banks (DB) are important.

- *Modern dive computers can process even the most complex biophysical models for diving?* Quite so these days. Chip speeds in modern dive computers rival Smartphones and range in the 800 $megaflops$ category (800 million flops) with $flops$ designating a floating point operation (add, subtract, multiply or divide) per sec. Lightning fast as that is and by contrast the fastest supercomputers today operate in the $pedaflops$ range, some million times faster. The Blue Mountain supercomputer at LANL was used to process the 3200 + downloaded diver profiles in the LANL DB to correlate and tune the USN, ZHL, VPM and RGBM algorithms and still took 13 min of wall clock time. Quantum computers on the horizon will be able to process that same task in roughly 2 sec or so.

- *Dive computers and dive planning software are supplanting traditional Dive Tables?* Yes again for the trained and experienced diver. Novices still use and likely benefit from learning and understanding Dive Tables, especially in recreational air and nitrox diving. In the technical arena, there are few Dive Tables with the NAUI Technical Dive Tables a singular exception. Check them out and use them safely.

- *All models are wrong but some are useful?* Boy, when is comes to computer, model, table and software fabrication there is a certain truth in the statement. A better declarative might be *incomplete* rather than *wrong* but the point is well taken in the diving arena. Dick Vann once remarked that "diving models are like socks – everybody has two and they both stink". Useful models, stinking or not, are ones that are correlated with real diving data, are trustworthy and are reproducible under a wide variety of environmental and physiological conditions. Not an easy nor trivial requirement for sure.

- *The original USN Tables were thoroughly tested?* Not even close. Ed Lanphier who was in charge of the program back in the 1950s reported at The AAUS Repetitive Dive Workshop at Woods Hole back in the 1990s that the USN Repetitive Dive Tables were "only tested for a few repetitive profiles" and the rest were "extrapolations". Luckily the USN model (Workman) had much conservatism built in because the slowest tissue compartment (120 min) was used to control repetitive diving.

- *Chamber and wet tests of a specific profile are not conclusive for all diving on all systems on all gas mixtures?* For sure but can often provide metrics for safe staging protocols whether successful or not. Here is where DBs shine because of the diversity of profiles and outcomes. And low cost versus chamber and wet tests.

- *Training Agencies are, or have been, conducting some of their own testing of shallow and deep stop models*? Quite so. PADI back 20 *years* or so conducted open water testing on the USN Tables for repetitive diving and came up with another set called the DSAT Tables. DSAT Tables have Spencer NDLs (more conservative than USN NDLs) and repetitive procedures less conservative than the original USN Tables. ANDI actively tested the RGBM using their Instructors before releasing new training standards and software (ANDI GAP) for air and nitrox diving. NAUI of course tested the RGBM over many years before extending training standards and issuing a full set of Technical Dive Tables for OC air, nitrox, helitrox and trimix plus RB Tables for some standard diluents. NAUI GAP software was released in the early 2000s tracking released NAUI Technical Dive Tables and more recently, and timely, Free Phase RGBM Simulator was packaged for sale both commercially and through NAUI Headquarters. Free Phase RGBM Simulator matches the NAUI Technical Dive Tables too. This is important for safety, training and necessary uniformity and reproducibility. Training by all Agencies has been efficient, safe and noteworthy whether using shallow or deep stop protocols.
- *Isobaric counterdiffusion (ICD) is just a theoretical concoction and never observed*? Nada. It's been demonstrated and observed in laboratory experiments with helium and nitrogen. Improper gas switches from nitrogen based mixes to helium based mixes (heavy-to-light) have been implicated as DCS causative. Simple calculations with diveware packages will easily quantify the effects on just dissolved gases.
- *The groupless, no-calculation recreational NAUI RGBM Tables are conservative*? Yes again and very easy to use versus the older versions of the NAUI Tables based on the USN model. A simple set of repetitive dives out to the limits of the NAUI RGBM Air and Nitrox Tables compared to allowable times in the old NAUI (USN) Tables will underscore the conservancy. For teaching Tables, the NAUI RGBM Air and Nitrox Tables are optimal for neophytes and useful for experienced divers.
- *The DCS hit rate in chamber tests is lower than open ocean tests for the same profile*? Another yes as noted by Peter Bennett. The variability of environmental impacts, diver comfort and awareness, buoyancy modifications and equipment demands in open water versus the tranquility and comfort of chamber and pod beds weigh more heavily on open water tests.
- *All Training Agencies teach Dive Tables*? Not anymore. Some mandate and teach dive computers for ease and simplicity as mainline policy. Guess it's sort of like using calculators for arithmetic instead of knowing how to add, subtract, multiply and divide. Your call not ours.
- *GFs have been tested and correlated with VPM and RGBM*? Not yet but that is an exercise that might well be a worthwhile undertaking to provide reference metrics against consistent, uniform and extensively used dive computer and software algorithms. GFs are arbitrary and not self consistent. Correlating GFs with tried, tested and safe BMs has been on a bucket list here at LANL for some time. We'll see.

- *Deep stops control the bubble and shallow stops treat the bubble?* Certainly true from a pure physics point of view but biochemistry and metabolic processes affect the tissue and blood in ways that have not been quantified nor incorporated into BMs (nor GMs). On the face of it, the statement (attributed to some in the medical community) is a simple description of staging differences between BMs and GMs. For arbitrary dives, equal surfacing risk BM protocols are always shorter then corresponding GM protocols. That's noteworthy for whatever reasons.
- *God created helium for diving but the Devil replaced it with nitrogen?* Just could be and if so Satan also made helium very expensive.

Fun and safe diving and hope this monogram contributes to it.

An Overview of
One Theory of DCS Physiology

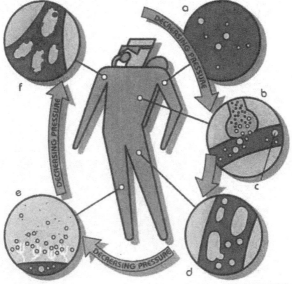

RON SYMES ARTWORK

a. As a diver ascends, nitrogen diffuses into microbubbles. Some of these microbubbles are transported through the heart and into the capillary beds of the lungs.

b. Once in the capillary beds, these bubbles are trapped. The gasses in these bubbles expand and leave the body through normal respiration. If the bubbles are not numerous, the lungs will clear them out sufficiently to prevent any occurrence of DCS.

c. Bubbles will form if there is too much nitrogen in a scuba diver's tissues — or if the diver ascends too rapidly. Too many bubbles may not be absorbed and could remain in the circulatory system.

d. As bubbles grow, they become elongated, and smaller bubbles begin diffusing into larger bubbles.

e. In the meantime microscopic bubbles of gas in ligaments and tendons can attract escaping nitrogen, which results in extravascular bubbles. If large enough this type of bubble can crowd and pinch nerves, giving the classic joint pain so typical of DCS.

f. Back in the circulatory system, nitrogen bubbles grow and attract blood platelets, constricting blood vessels. Protein is released which causes the blood to "sludge," or thicken, and blood volume drops.

Keep in mind that symptomatic bubbles do not necessarily form on every dive. This illustration represents what could happen to a diver who has either stayed at depth too long, ascended too rapidly, or incurred DCS for a more subtle physiological reason.

Applications and Exercises

A first set of exercises poses problems with solutions and is included as a warmup for the next set directed at dive computer applications. These first ones ought tweak your general diving saavy too. Have fun with both.

1. How many nautical miles to a kilometer?

$$1\,nautical\,mile = 1.85\,km\ ,\quad 1\,km = \frac{1}{1.85}$$
$$nautical\,mile = 0.54\,nautical\,mile$$

2. How many electrostatic units (esu) to a coulomb?

$$1\,coul = 2.99 \times 10^9\,esu\ ,\quad 1\,esu = \frac{1}{2.99 \times 10^9}\,coul = 3.34 \times 10^{-10}\,coul$$

3. How many light years to a mile?

$$1\,light\,yr = 5.88 \times 10^{12}\,mile\ ,\quad 1\,mile$$
$$= \frac{1}{5.88 \times 10^{12}}\,light\,yr = 1.70 \times 10^{-13}\,light\,yr$$

4. Convert depth, $d = 38\,fsw$, to depth, ffw, in fresh water?

$$38\,fsw \times \frac{1\,ffw}{0.975\,fsw} = 38.9\,ffw$$

B. R. Wienke, T. R. O'Leary, *Understanding Modern Dive Computers and Operation*, SpringerBriefs in Computer Science, https://doi.org/10.1007/978-3-319-94054-0_11

5. Convert ascent rate, $r = 60 \, fsw/min$, to msw/sec?

$$r = 60 \, fsw/min \times \frac{msw}{3.28 \, fsw} \times \frac{min}{60 \, sec} = 0.305 \, msw/sec$$

6. Convert volume, $V = 6.2 \, m^3$, to ft^3?

$$V = 6.2 \, m^3 \times \frac{353.2 \, ft^3}{m^3} = 2189 \, ft^3$$

7. Convert pressure, $P = 5.3 \, kg/m^2$, to lb/in^2?

$$P = 5.3 \, kg/m^2 \times \frac{0.20 \, lb/ft^2}{1 \, kg/m^2} \times \frac{1 \, ft^2}{144 \, in^2} = 0.0074 \, lb/in^2$$

8. Convert acceleration, $g = 32 \, ft/sec^2$, to m/sec^2?

$$g = 32 \, ft/sec^2 \times \frac{1 \, m}{3.28 \, ft} = 9.8 \, m/sec^2$$

9. What is the specific density, η, of mercury (Hg) with respect to seawater?

$$\rho_{Hg} = 13.55 \, g/cm^3 \quad , \quad \rho_{seawater} = 1.026 \, gm/cm^3$$

$$\eta = \frac{\rho_{Hg}}{\rho_{seawater}} = \frac{13.55}{1.026} = 13.21$$

10. A freely falling body has instantaneous (vertical) trajectory, y, given by the relationship,

$$y = y_i + v_i t - \frac{1}{2} g t^2$$

with t the time, y_i the initial position, v_i the initial velocity and g the acceleration of gravity. What is the instantaneous velocity, v, and instantaneous acceleration, a?

$$y = y_i + v_i t - \frac{1}{2} g t^2 \quad , \quad v = \frac{dy}{dt} \quad , \quad a = \frac{dv}{dt}$$

$$v = \frac{d(y_i + v_i t - 1/2 g t^2)}{dt} = v_i - gt$$

$$a = \frac{d(v_i - gt)}{dt} = -g$$

11. Convert $37\,^{\circ}C$ to Fahrenheit $(^{\circ}F)$ and then to Rankine $(^{\circ}R)$ temperatures?

$$^{\circ}F = \frac{9}{5}\,^{\circ}C + 32 = \frac{9}{5} \times 37 + 32 = 98.6^{\circ}$$

$$^{\circ}R =^{\circ} F + 460 = 98.6 + 460 = 558.6^{\circ}$$

12. Convert $80\,^{\circ}F$ to Centigrade $(^{\circ}C)$ and then to Kelvin $(^{\circ}K)$ temperatures?

$$^{\circ}C = \frac{5}{9}\,(^{\circ}F - 32) = \frac{5}{9}\,(80 - 32) = 26.6^{\circ}$$

$$^{\circ}K =^{\circ} C + 273 = 26.6 + 273 = 299.6^{\circ}$$

13. What volume, V, does a *gmole* of an ideal gas occupy at standard temperature and pressure?

$$P = 10.1\ newton/cm^2\ , \quad T = 273\,^{\circ}K\ , \quad R = 8.317\ j/gmole$$

$$PV = nRT\ , \quad V = \frac{nRT}{P}$$

$$V = \frac{8.317 \times 273}{0.101}\ cm^3 = 22.48 \times 10^3\ cm^3 = 22.48\ l$$

14. If the jet stream is traveling, v, at $155\ mi/hr$, how long, t, does it take to cross the USA assuming a straight line flow pattern and width across USA, l, equal to $2{,}600\ mi$?

$$t = \frac{l}{v} = \frac{2{,}600}{155}\ hr = 16.7\ hr$$

15. If the jet stream wobbles into circular subflows with radii, r, approximately $200\ mi$ from center of the flow, what is the centripetal acceleration, a, experienced by the jet stream?

$$a = \frac{v^2}{r} = \frac{155^2}{200}\ mi/hr^2 = \frac{24025}{200}\ mi/hr^2 = 120.1\ mi/hr^2$$

16. A $448\ lb$ winch gear, displacing a water volume, $V = 2\ ft^3$, rests on a hard sea bottom at $99\ fsw$. What surface volume of air, V_{sur}, is needed to inflate lift bags to bring the gear to the surface?

$$d = 99\ fsw\ , \quad \rho = 64\ lbs/ft^3\ , \quad w = 448\ lbs$$

$$V_{lift} = \frac{w}{\rho} = \frac{448}{64} \, ft^3 = 7 \, ft^3$$

$$V_{sur} = V_{lift} \left[1 + \frac{d}{33} \right] = 4 \times 7 \, ft^3 = 28 \, ft^3$$

17. A buoy weighing 48 *lbs* occupies, $V = 3 \, ft^3$. What fraction, ξ, of its volume will float above water?

$$V = 3 \, ft^3 \, , \quad \xi = \frac{V - V_{dis}}{V}$$

$$V_{dis} = \frac{w}{\rho} = \frac{48}{64} \, ft^3 = 0.75 \, ft^3$$

$$\xi = \frac{3 - 0.75}{3} = 0.75$$

18. What are composite partial pressures, p_i, for a TMX 80/10 breathing gas at ocean depth of 400 fsw?

$$p_i = f_i \, (33 + 400) \, fsw \quad (i = He, O_2, N_2)$$

$$p_{He} = 0.80 \times 433 \, fsw = 346.4 \, fsw$$

$$p_{O_2} = 0.10 \times 433 \, fsw = 43.3 \, fsw$$

$$p_{N_2} = 0.10 \times 433 \, fsw = 43.3 \, fsw$$

19. In gel experiments, if the critical radius, ϵ, is inversely proportional to the crushing pressure differential, ΔP, what happens to the critical radius if the crushing differential is tripled?

$$\Delta P_f \epsilon_f = \Delta P_i \epsilon_i \, , \quad \Delta P_f = 3 \Delta P_i$$

$$\epsilon_f = \frac{\epsilon_i}{3}$$

20. Laboratory bubble seed counts in gels and (some) living tissue confirm the seed size (radius), r, distribution, n, is exponential, decreasing in number as the seed radius increases, so that (differentially),

$$n_i = n_0 \, exp \, (-\beta r_c)$$

with n_0 and β constants. For small sample counts (microscope), $n_1 = 9865$ $r_1 = 0.7 \, microns$ and $n_2 = 5743$, $r_2 = 1.4 \, microns$, what are n_0 and β?

$$n_i = n_0 \, exp \, (-\beta r_i) \, , \quad \ln \, (n_1/n_2) = -\beta(r_1 - r_2)$$

$$\beta = \frac{1}{r_2 - r_1} \ln \, (n_1/n_2) = \frac{1}{0.7} \ln \, (9865/5743) = 0.773$$

$$n_0 = n_i \, exp \, (\beta r_c) = n_1 \, exp \, (\beta r_1) = 9865 \, exp \, (0.773 \times 0.7) = 16947$$

Assuming β is determined (given), how is the distribution function, n, normalized to the total seed count, N, across all sizes?

$$n_0 \int_0^\infty exp \, (-\beta r) \, dr = \frac{n_0}{\beta} = N$$

$$n_0 = \beta N$$

21. The air pressure in a scuba tank drops from 2475 lbs/in^2 to 1500 lbs/in^2 in 8 min. What is the air consumption rate, χ?

$$\chi = \frac{2475 - 1500}{8} \, lbs/in^2 \, min = 121.9 \, lbs/in^2 \, min$$

If the tank is rated at 72 ft^3, what is the consumption rate, χ, in ft^3/min?

$$121.9 \, lbs/in^2 \, min \times \frac{72 \, ft^3}{2475 \, lbs/in^2} = 3.5 \, ft^3/min$$

22. How long, t, will a tank containing, $V = 34 \, ft^3$, of air last at 33 fsw for an EOD specialist swimming against a 6 $knot$ very cold current in the ocean?

$$P_0 = 33 \, fsw \, , \quad \chi_0 = 2 \, ft^3/min \, , \quad \chi = \chi_0 \left[1 + \frac{d}{P_0} \right]$$

$$\chi = 2 \times \left[1 + \frac{33}{33} \right] \, ft^3/min = 4 \, ft^3/min$$

$$t = \frac{V}{\chi} = \frac{34}{4} \, min = 8.5 \, min$$

23. A CCR booster rated 80 ft^3 at 3000 lb/in^2, registers a pressure, $P = 1420 \, lb/in^2$ on a sub gauge. What is the remaining air volume, V?

$$V = V_r \frac{P}{P_r}$$

$$V = 80 \times \frac{1420}{3000} \, ft^3 = 37.8 \, ft^3$$

What is the CCR booster constant, κ?

$$\kappa = \frac{P_r}{V_r} = \frac{3000}{80} \, lb/in^2 \, ft^3 = 37.5 \, lb/in^2 \, ft^3$$

24. According to Graham, what roughly is the ratio, ψ, of molecular diffusion speeds of hydrogen to oxygen?

$$\psi = \left[\frac{A_{O_2}}{A_{H_2}}\right]^{1/2} = \left[\frac{32}{2}\right]^{1/2} = 4$$

25. A commercial diving operation is constructing a set of helium proprietary tables using the popular DCIEM nitrogen tables as a basis before testing. If the spectrum of tissues, τ, in the DCIEM nitrogen tables is (2.5, 5, 10, 20, 40, 80, 160, 320 min), what are the corresponding set for the helium tables assuming the same critical tensions, M, as the nitrogen tables?

$$\tau_{He} = \left[\frac{A_{He}}{A_{N_2}}\right]^{1/2} \tau_{N_2} = \left[\frac{4}{28}\right]^{1/2} \tau_{N_2} = 0.38 \times \tau_{N_2}$$

$$\tau_{He} = (.94, 1.89, 3.78, 7.56, 15.12, 30.24, 60.48, 120.46) \, min$$

What might be a more convenient set of tissue halftimes?

$$\tau_{He} = (1, 2, 4, 8, 15, 30, 60, 120)$$

26. What is the optimal diving mixture for a decompression dive to 300 fsw holding maximum oxygen partial pressure, $pp_{O_2} = 1.2 \, atm$ and maximum nitrogen partial pressure, $p_{N_2} = 3.2 \, atm$ in a fresh water lake at 2,300 ft in the mountains?

$$\alpha = 0.038 \ , \ \ p_{N_2} = 3.2 \, atm \ , \ \ pp_{O_2} = 1.2 \, atm$$

$$\eta = 0.975 \ , \ \ h = 2.3 \ , \ \ d = 300 \, fsw \ , \ \ P_h = 33 \, exp$$

$$(-0.038 \times 2.3) \, fsw = 30.2 \, fsw$$

$$fo_2 = \frac{33 pp_{O_2}}{\eta d + P_h} = \frac{33 \times 1.2}{0.975 \times 300 + 30.2} = 0.123$$

$$f_{N_2} = \frac{33 P_{N_2}}{\eta d + P_h} = \frac{33 \times 1.2}{0.975 \times 300 + 30.2} == 0.328$$

$$f_{He} = 1 - fo_2 - f_{N_2} = 1 - 0.123 - 0.328 = 0.549$$

27. What is the surfacing oxygen partial pressure, p_0, for a normoxic breathing mixture at 450 fsw?

$$p = 0.21 \, atm \; (normoxic) \; , \quad P_0 = 33 \, fsw \; , \quad P = 450+33 \, fsw = 483 \, fsw$$

$$p_0 = \frac{P_0}{P} \, p = \frac{33}{483} \times 0.2 \, atm = 0.0137 \, atm$$

What can you say about such a mixture at the surface?

$$p_0 \leq 0.16 \, atm$$

$$Mixture \; Is \; Hypoxic \; (Very \; Hypoxic)$$

28. Assuming surface equilibration on air, what is the total tissue tension, Π, in the, $\tau = 20 \, min$, compartment after $10 \, min$ at depth, $d = 90 \, fsw$, for a salvage diver breathing TMX 60/25? $f_{O_2} = 0.15$)?

$$\Pi = p_{He} + p_{N_2} \; , \quad d = 90 \, fsw \; , \quad \tau_{N_2} = 20 \, min \; , \quad \tau_{He} = \frac{20}{2.65} = 7.55 \, min$$

$$\lambda_{N_2} = \frac{0.693}{\tau_{N_2}} = \frac{0.693}{20} \, min^{-1} = 0.0347 \, min^{-1}$$

$$\lambda_{He} = \frac{0.693}{\tau_{He}} = \frac{0.693}{7.55} \, min^{-1} = 0.0918 \, min^{-1}$$

$$p_{aN_2} = f_{N_2} p_a = f_{N_2}(33+d) \, fsw \; , \quad p_{iN_2} = 0.79 P_0$$

$$p_{aHe} = f_{He} p_a = f_{He}(33+d) \, fsw \; , \quad p_{iHe} = 0.0 \, fsw$$

$$p_{N_2} = p_{aN_2} + (p_{iN_2} - p_{aN_2}) \, exp \, (-\lambda_{N_2} t)$$

$$p_{He} = p_{aHe} + (p_{iHe} - p_{aHe}) \, exp \, (-\lambda_{He} t)$$

$$p_{iN_2} = 0.79 \times 33 \, fsw = 26.01 \, fsw \; ,$$

$$p_{aN_2} = f_{N_2} p_a = 0.25 \times 123 = 30.7 \, fsw$$

$$p_{N_2} = 30.7 + (26.1 - 30.7) \, exp \, (-0.0347 \times 10) \, fsw = 27.4 \, fsw$$

$$p_{iHe} = 0.0 \, fsw \; , \quad p_{aHe} = f_{He} p_a = 0.60 \times 123 \, fsw = 73.8 \, fsw$$

$$p_{He} = 73.8 - 73.8 \, exp \, (-0.0918 \times 10) \, fsw = 44.3 \, fsw$$

$$\Pi = 27.4 + 44.3 \, fsw = 71.7 \, fsw$$

What is the USN critical surfacing tension, M_0, for the 20 min compartment?

$$M_0 = 72 \, fsw$$

Can this diver ascend to the surface on his trimix?

$$Probably \; - \; But \; Slowly$$

29. How long does it take for the 80 min compartment to approach its USN critical surfacing tension, $M = M_0 = 52 \, fsw$, at depth of 140 fsw assuming initial nitrogen tension of 45 fsw?

$$p_i = 45 \, fsw \; , \quad p_a = f_{N_2}(33 + d)$$

$$p_a = 0.79 \times (33 + 140) \, fsw = 136.6 \, fsw$$

$$\lambda = \frac{0.693}{80} \, min^{-1} = 0.0087 \, min \; , \quad M = 52 \, fsw$$

$$t = \frac{1}{\lambda} \ln \left[\frac{p_i - p_a}{M - p_a} \right] = 114.9 \times \ln \left[\frac{91.6}{84.6} \right] \, min = 9.1 \, min$$

What is the nonstop limit, t_n, for the 80 min tissue at this depth?

$$t_n = 9.1 \, min$$

30. If the nonstop time limit at depth, $d = 90 \, fsw$, is, $t_n = 22 \, min$, what is the surfacing critical tension, M_0, assuming that the 5 min compartment controls the exposure (has largest computed tissue tension at this depth)?

$$\lambda = \frac{0.693}{5} \, min^{-1} = 0.1386 \, min^{-1}$$

$$p_i = 0.79 \times 33 \, fsw = 26.1 \, fsw$$

$$p_a = 0.79 \times (33 + 90) = 97.1 \, fsw$$

$$M_0 = p_a + (p_i - p_a) \, exp \, (-\lambda t_n)$$

$$M_0 = 97.1 - 78.2 \, exp \, (-0.1386 \times 22) \, fsw = 94 \, fsw$$

The following exercises mimic dive computers, drop down menus, sensors, interfaces and numerical operations before, during and after a dive. Hopefully they provide additional insight into the foregoing discourse and development. And hopefully the foregoing set has given you all the information you need to tackle these. Answers are in boldface.

1. What are two short descriptors for GM and BM dive computers – *one controls dissolved gas and the other focuses on bubbles?*

   (a) **shallow stop; deep stop computers**.
   (b) RB; OC computers.
   (c) recreational; technical computers.

2. Nominal ascent rates in dive computers are ——— *results from Doppler bubble counting in recreational divers?*

   (a) $10\,fsw/min$.
   (b) **$30\,fsw/min$**.
   (c) $60\,fsw/min$.

3. USN and ZHL models are ——— algorithms while VPM and RGBM models are ——— algorithms.

   (a) GM, Haldane.
   (b) **GM, BM**.
   (c) BM, Yount.

4. A CCR booster rated $80\,ft^3$ at $3000\,lb/in^2$, registers a pressure, $P = 1420\,lb/in^2$ on a sub gauge so what is the remaining booster gas volume, V – *recall your tank pressure-volume relationships and tank constant definition?*

   (a) **$37.8\,ft^3$**.
   (b) $80\,ft^3$.
   (c) $23.6\,ft^3$.

5. Shallow safety stops are made ——— *what the name suggests?*

   (a) at the surface.
   (b) **in the 10–20 fsw zone**.
   (c) at 1/2 the bottom depth.

6. Deep stops are made ——— *recall bubble models?*

   (a) **consistent with bubble dynamics**.
   (b) to minimize dissolved gas elimination.
   (c) at 1/2 the depth of the first decompression stop.

7. How much fresh water, V, does a $200\,lb$ lift bag displace – *remember Archimedes?*

   (a) $6.4\,ft^3$.
   (b) $9.6\,ft^3$.
   (c) **$3.2\,ft^3$**

8. A fully inflated BC displaces, $V = 0.78\,ft^3$, of sea water. What is the lift, B, provided by the BC – *Archimedes again?*

   (a) $40.4\,lbs$.
   (b) **$49.9\,lbs$**.
   (c) $48.6\,lbs$.

9. What is the tissue tension, p, in the $80\,min$ compartment of an air diver at $90\,fsw$ for 20 min and first equilibrated at sea level before the dive – *meaning* $p_i = 0.79 \times 33$ *fsw*, $p_a = 0.79 \times 123$ *fsw*, $\tau = 80$ *min and nitrogen fraction of air is 0.79, of course, and you can do this by hand without diveware?*

   (a) $26\,fsw$.
   (b) $97\,fsw$.
   (c) **37 fsw**

10. If heliair with helium fraction 0.79 were substituted for the same dive, what would be the helium tension, p, in the same compartment – *same as above assuming switch to helium mix off equilibrated air at the surface, so that* $p_i = 0\,fsw$ *for helium this time at the surface?*

   (a) $90\,fsw$.
   (b) **39 fsw**.
   (c) $33\,fsw$.

11. For the same helium dive, what is the tissue tension of air in the $80\,min$ compartment – *tricky, so* $p_i = 0.79 \times 33$ *fsw still but* $p_a = 0$ *fsw for air at the bottom of the heliair dive?*

   (a) $123\,fsw$.
   (b) $39\,fsw$.
   (c) **23 fsw**.

12. According to an air GM computer, what is the surfacing M-value in the $40\,min$ compartment and what might a diver with inert gas (nitrogen plus helium) tension 70 fsw – *just a little M-value arithmetic with the equation?*

   (a) $61\,fsw$; proceed directly to the surface.
   (b) $72\,fsw$; make a decompression stop.
   (c) **61 fsw; make a decompression stop**.

13. For an air depth-nonstop time law of the form, $dt_n^{1/2} = C$, what is the nonstop time, t_n, limit for compartment, $\tau = 45\,min$, and what is the depth, d, for $C = 450\,fsw\,min^{1/2}$ – *first recall the halftime-nonstop relationship,* $\lambda t_n = 1.25$?

   (a) **81 min, 49 fsw**.
   (b) $67\,min$, $38\,fsw$.
   (c) $81\,min$, $56\,fsw$.

14. Audible and displayed computer warnings sent to divers include – *standard protocol across dive computer Manufacturers?*

   (a) **ascent rate violations, OT violations, missed stops**.
   (b) ascent rate violations, unsafe gas mixtures, breathing loop tears.
   (c) air consumption violations, OT violations, breathing loop tears.

15. If the air nonstop time limit at depth, $d = 90\,fsw$, is $t_n = 22\,min$, what is the surfacing critical tension, M_0, assuming that the $5\,min$ compartment controls the exposure (has largest computed tissue tension at this depth) – *just invert the tissue equation for tension equal M_0 at that time?*

(a) $M_0 = 43\,fsw$.
(b) $M_0 = 103\,fsw$.
(c) $M_0 = 94\,\textbf{fsw}$

16. If the separated phase volume calculated by a BM computer is $250\,microns^3$ at $66\,fsw$, what will be the surfacing value and can a mixed gas diver ascend directly to the surface – *remembering Boyle's pressure-volume relationship?*

(a) $250\,microns^3$; no.
(b) $750\,microns^3$; yes.
(c) $\textbf{750\,microns}^3\textbf{; no.}$

17. What is the surfacing bubble volume, ϕ, in the $13.3\,min$ helium tissue compartment excited into growth on a heliox (80/20) dive to $300\,fsw$ for $20\,min$ neglecting ascent and descent time assuming surface heliox equilibration for simplicity and using the given constants, excitation radius, $r_{ex} = 1\,micron$ in unit biomass, $DS = 4\,micron^3/fsw$, $N = 1$, $2\,\gamma = 1\,fsw\,micron$ and $\beta = 1\,micron^{-1}$ – *evaluate integral directly with the unit bubble density and helium tension at $20\,min$. MATHEMATICA will help on this one?*

(a) $4349\,microns^3$.
(b) $\textbf{772\,microns}^3$.
(c) $434.9\,microns^3$.

18. What are the pulmonary, Γ, and CNS, Ω, toxicities on an air dive to $130\,fsw$ for $25\,min$ – *just straightforward use of toxicity equations?*

(a) 0.099; $80.3\,min$;
(b) 0.456; $80.3\,min$.
(c) $\textbf{0.099; 26.3\,min.}$

19. If the nitrogen halftime, τ_{N_2}, in the fastest compartment is $3\,min$, what is the corresponding helium halftime, τ_{He} – *trivial, we hope?*

(a) $\textbf{1\,min.}$
(b) $3\,min.$
(c) $9\,min.$

20. What is the instantaneous risk, r_{GM}, surfacing from a nonstop air dive to $130\,fsw$ for $4\,min$ neglecting possible outgassing on ascent and the $8\,min$ tissue compartment controlling the ascent – *use GM risk equations with usual gas loadings and Buhlmann Z-values?*

(a) 0.024
(b) 0.128
(c) $\textbf{0.083}$

21. What is the instantaneous risk, r_{BM}, surfacing from a TMX 16/40 dive to 240 fsw for 15 min with the 2.4 min compartment controlling at the bottom, permissible bubble supersaturation, $H = 20\,fsw$ and neglecting possible outgassing and bubble growth on emergency ascent – *use BM risk equations with tissue equations but better yet get Free Phase Simulator, GAP, RGBM Simulator or any of the software packages offered by BM computer Vendors?*

    (a) **0.464**
    (b) 0.401
    (c) 0.133

22. What is the surfacing tension in the controlling 3.5 min tissue compartment after a TMX 16/40 dive to 300 fsw for 20 min with emergency ascent rate of 30 fsw/min – *use tissue equations with ascent rate included or again get Free Phase Simulator, GAP, RGBM Simulator or any of the software packages offered by BM computer Vendors?*

    (a) 0.018
    (b) **0.416**
    (c) 0.194

What is the surfacing risk, r_{BM}, neglecting outgassing on the way up – *again BM risk equations or BM software?*

    (a) **0.524**
    (b) 0.639
    (c) 0.081

What is the surfacing risk, r_{GM}, including outgassing on the way up – *use GM risk equations plus Buhlmann Z-values and tissue equations with ascent or get any GM software package?*

    (a) 0.757
    (b) 0.431
    (a) **0.340**

23. What is ambient pressure, p_a, for a diver ascending at constant acceleration – *recall Newton's law of displacement, $1/2at^2$, with a in the same partial pressure units as p and p_0?*

    (a) $p_a = p_0 - at$.
    (b) $\mathbf{p_a = p_0 - 1/2at^2}$.
    (c) $p_a = vt - at$.

What is the tissue equation with constant acceleration, a, upward – *formulate p_a with constant change due to acceleration and see the constant velocity case in the text?*

(a) $\partial \mathbf{p}/\partial \mathbf{t} + \lambda \mathbf{p} = \lambda(\mathbf{p_0} - 1/2\mathbf{a}\mathbf{t}^2)$.
(b) $\partial p/\partial t + \lambda p = \lambda(p_0 + 1/2at^2)$.
(c) $\partial p/\partial t + \lambda p = -\lambda(p_0 - 1/2at^2)$

What is the solution to the tissue equation with constant acceleration – *turn the crank for a linear first order differential equation with quadratic dependence on time, t?*

(a) $p = p_0 + at/\lambda - a/\lambda^2 + (a/\lambda^2 + p_0)exp\,(-\lambda t) + 1/2at^2$.
(b) $p = p_0 - at/\lambda - a/\lambda^2 + (a/\lambda^2 + p_i)exp\,(-\lambda t) + 1/2at^2$.
(c) $\mathbf{p} = \mathbf{p_0} + \mathbf{at}/\lambda - \mathbf{a}/\lambda^2 + (\mathbf{a}/\lambda^2 + \mathbf{p_i} - \mathbf{p_0})\mathbf{exp}\,(-\lambda \mathbf{t}) - 1/2\mathbf{a}\mathbf{t}^2$.

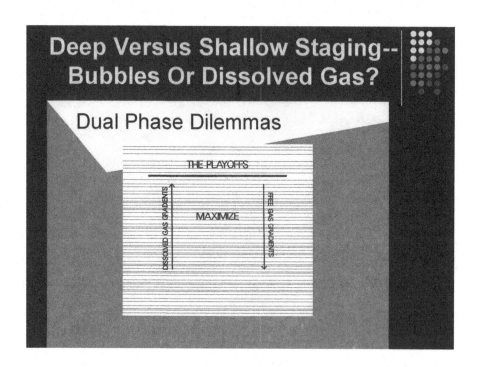

"God invented helium for diving but the devil replaced it with nitrogen"

"Deep stops control the bubble and shallow stops treat the bubble"

"All models are wrong but some are useful"

C&C Dive Team
Prep Room
1987

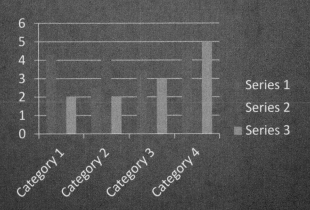

Printed in the United States
By Bookmasters

Printed in the United States
By Bookmasters